Connected  Mathematics

Data About Us

Statistics

Teacher's Edition

Glenda Lappan
James T. Fey
William M. Fitzgerald
Susan N. Friel
Elizabeth Difanis Phillips

Developed at Michigan State University

DALE SEYMOUR PUBLICATIONS®

The Connected Mathematics Project was developed at Michigan State University with financial support from the Michigan State University Office of the Provost, Computing and Technology, and the College of Natural Science.

This material is based upon work supported by the National Science Foundation under Grant No. MDR 9150217.

This project was supported, in part,
by the
National Science Foundation
Opinions expressed are those of the authors
and not necessarily those of the Foundation

The Michigan State University authors and administration have agreed that all MSU royalties arising from this publication will be devoted to purposes supported by the Department of Mathematics and the MSU Mathematics Education Enrichment Fund.

This book is published by Dale Seymour Publications®, an imprint of the Alternative Publishing Group of Addison-Wesley Publishing Company.

Managing Editor: Catherine Anderson
Project Editor: Stacey Miceli
Book Editor: Mali Apple
Production/Manufacturing Director: Janet Yearian
Production/Manufacturing Coordinator: Claire Flaherty
Design Manager: John F. Kelly
Photo Editor: Roberta Spieckerman
Design: Don Taka
Composition: London Road Design, Palo Alto, CA
Illustrations: Pauline Phung, Margaret Copeland, Ray Godfrey
Cover: Ray Godfrey

Photo Acknowledgements: 6 © John Griffin/The Image Works; 9 © John Whitmer/Stock, Boston; 23 (left) © Rhoda Sidney/The Image Works; 23 (right) © James Fosset/The Image Works; 26 © Ulf Sjostedt/FPG International; 28 © Gale Zucker/Stock, Boston; 30 © Michael Dwyer/Stock, Boston; 53 (top right) © Spencer Grant/Photo Researchers, Inc.; 53 (lower left) © Mike Valeri/FPG International; 53 (lower right) © Lionel Delevingne/Stock, Boston; 59 (left) © Mike Douglas/The Image Works; 59 (right) © Jonathan Meyers/FPG International; 63 © Michael Dwyer/Stock, Boston

Yahtzee is a trademark of the Milton Bradley Company.

DALE SEYMOUR PUBLICATIONS®
P.O. BOX 10888
PALO ALTO, CA 94303

Order number 21444
ISBN 1-57232-149-0

1 2 3 4 5 6 7 8 9 10-ML-99 98 97 96 95

The Connected Mathematics Project Staff

Project Directors

James T. Fey
University of Maryland

William M. Fitzgerald
Michigan State University

Susan N. Friel
University of North Carolina at Chapel Hill

Glenda Lappan
Michigan State University

Elizabeth Difanis Phillips
Michigan State University

Project Manager

Kathy Burgis
Michigan State University

Technical Coordinator

Judith Martus Miller
Michigan State University

Collaborating Teachers/Writers

Mary K. Bouck
Portland, Michigan

Jacqueline Stewart
Okemos, Michigan

Curriculum Development Consultants

David Ben-Chaim
Weizmann Institute

Alex Friedlander
Weizmann Institute

Eleanor Geiger
University of Maryland

Jane Mitchell
University of North Carolina at Chapel Hill

Anthony D. Rickard
Alma College

Evaluation Team

Diane V. Lambdin
Indiana University

Sandra K. Wilcox
Michigan State University

Judith S. Zawojewski
National-Louis University

Graduate Assistants

Scott J. Baldridge
Michigan State University

Angie S. Eshelman
Michigan State University

M. Faaiz Gierdien
Michigan State University

Jane M. Keiser
Indiana University

Angela S. Krebs
Michigan State University

James M. Larson
Michigan State University

Ronald Preston
Indiana University

Tat Ming Sze
Michigan State University

Sarah Theule-Lubienski
Michigan State University

Jeffrey J. Wanko
Michigan State University

Field Test Production Team

Katherine Oesterle
Michigan State University

Stacey L. Otto
University of North Carolina at Chapel Hill

Teacher/Assessment Team

Kathy Booth
Waverly, Michigan

Anita Clark
Marshall, Michigan

Theodore Gardella
Bloomfield Hills, Michigan

Yvonne Grant
Portland, Michigan

Linda R. Lobue
Vista, California

Suzanne McGrath
Chula Vista, California

Nancy McIntyre
Troy, Michigan

Linda Walker
Tallahassee, Florida

Software Developer

Richard Burgis
East Lansing, Michigan

Development Center Directors

Nicholas Branca
San Diego State University

Dianne Briars
Pittsburgh Public Schools

Frances R. Curcio
New York University

Perry Lanier
Michigan State University

J. Michael Shaughnessy
Portland State University

Charles Vonder Embse
Central Michigan University

Field Test Coordinators

Michelle Bohan
Queens, New York

Melanie Branca
San Diego, California

Alecia Devantier
Shepherd, Michigan

Jenny Jorgensen
Flint, Michigan

Sandra Kralovec
Portland, Oregon

Sonia Marsalis
Flint, Michigan

William Schaeffer
Pittsburgh, Pennsylvania

Karma Vince
Toledo, Ohio

Virginia Wolf
Pittsburgh, Pennsylvania

Shirel Yaloz
Queens, New York

Student Assistants

Laura Hammond
David Roche
Courtney Stoner
Jovan Trpovski
Julie Valicenti
Michigan State University

Patricia Wagner
Holmes Middle School

Greg Williams
Gundry Elementary School

Lansing

Susan Bissonette
Waverly Middle School

Kathy Booth
Waverly East Intermediate School

Carole Campbell
Waverly East Intermediate School

Gary Gillespie
Waverly East Intermediate School

Denise Kehren
Waverly Middle School

Virginia Larson
Waverly East Intermediate School

Kelly Martin
Waverly Middle School

Laurie Metevier
Waverly East Intermediate School

Craig Paksi
Waverly East Intermediate School

Tony Pecoraro
Waverly Middle School

Helene Rewa
Waverly East Intermediate School

Arnold Stiefel
Waverly Middle School

Portland

Bill Carlton
Portland Middle School

Kathy Dole
Portland Middle School

Debby Flate
Portland Middle School

Yvonne Grant
Portland Middle School

Terry Keusch
Portland Middle School

John Manzini
Portland Middle School

Mary Parker
Portland Middle School

Scott Sandborn
Portland Middle School

Shepherd

Steve Brant
Shepherd Middle School

Mary Brock
Shepherd Middle School

Cathy Church
Shepherd Middle School

Ginny Crandall
Shepherd Middle School

Craig Ericksen
Shepherd Middle School

Natalie Hackney
Shepherd Middle School

Bill Hamilton
Shepherd Middle School

Julie Salisbury
Shepherd Middle School

Sturgis

Sandra Allen
Eastwood Elementary School

Margaret Baker
Eastwood Elementary School

Steven Baker
Eastwood Elementary School

Keith Barnes
Eastwood Elementary School

Wilodean Beckwith
Eastwood Elementary School

Darcy Bird
Eastwood Elementary School

Bill Dickey
Sturgis Middle School

Ellen Eisele
Eastwood Elementary School

James Hoelscher
Sturgis Middle School

Richard Nolan
Sturgis Middle School

J. Hunter Raiford
Sturgis Middle School

Cindy Sprowl
Eastwood Elementary School

Leslie Stewart
Eastwood Elementary School

Connie Sutton
Eastwood Elementary School

Traverse City

Maureen Bauer
Interlochen Elementary School

Ivanka Berskshire
East Junior High School

Sarah Boehm
Courtade Elementary School

Marilyn Conklin
Interlochen Elementary School

Nancy Crandall
Blair Elementary School

Fran Cullen
Courtade Elementary School

Eric Dreier
Old Mission Elementary School

Lisa Dzierwa
Cherry Knoll Elementary School

Ray Fouch
West Junior High School

Ed Hargis
Willow Hill Elementary School

Richard Henry
West Junior High School

Dessie Hughes
Cherry Knoll Elementary School

Ruthanne Kladder
Oak Park Elementary School

Bonnie Knapp
West Junior High School

Sue Laisure
Sabin Elementary School

Stan Malaski
Oak Park Elementary School

Jody Meyers
Sabin Elementary School

Marsha Myles
East Junior High School

Mary Beth O'Neil
Traverse Heights Elementary School

Jan Palkowski
East Junior High School

Karen Richardson
Old Mission Elementary School

Kristin Sak
Bertha Vos Elementary School

Mary Beth Schmitt
East Junior High School

Mike Schrotenboer
Norris Elementary School

Gail Smith
Willow Hill Elementary School

Karrie Tufts
Eastern Elementary School

Mike Wilson
East Junior High School

Tom Wilson
West Junior High School

Minnesota

Minneapolis

Betsy Ford
Northeast Middle School

New York

East Elmhurst

Allison Clark
Louis Armstrong Middle School

Dorothy Hershey
Louis Armstrong Middle School

J. Lewis McNeece
Louis Armstrong Middle School

Rossana Perez
Louis Armstrong Middle School

Merna Porter
Louis Armstrong Middle School

Marie Turini
Louis Armstrong Middle School

North Carolina

Durham

Everly Broadway
Durham Public Schools

Thomas Carson
Duke School for Children

Mary Hebrank
Duke School for Children

Bill O'Connor
Duke School for Children

Ruth Pershing
Duke School for Children

Peter Reichert
Duke School for Children

Elizabeth City

Rita Banks
Elizabeth City Middle School

Beth Chaundry
Elizabeth City Middle School

Amy Cuthbertson
Elizabeth City Middle School

Deni Dennison
Elizabeth City Middle School

Jean Gray
Elizabeth City Middle School

John McMenamin
Elizabeth City Middle School

Nicollette Nixon
Elizabeth City Middle School

Malinda Norfleet
Elizabeth City Middle School

Joyce O'Neal
Elizabeth City Middle School

Clevie Sawyer
Elizabeth City Middle School

Juanita Shannon
Elizabeth City Middle School

Terry Thorne
Elizabeth City Middle School

Rebecca Wardour
Elizabeth City Middle School

Leora Winslow
Elizabeth City Middle School

Franklinton

Susan Haywood
Franklinton Elementary School

Clyde Melton
Franklinton Elementary School

Louisburg

Lisa Anderson
Terrell Lane Middle School

Jackie Frazier
Terrell Lane Middle School

Pam Harris
Terrell Lane Middle School

Ohio

Toledo

Bonnie Bias
Hawkins Elementary School

Marsha Jackish
Hawkins Elementary School

Lee Jagodzinski
DeVeaux Junior High School

Norma J. King
Old Orchard Elementary School

Margaret McCready
Old Orchard Elementary School

Carmella Morton
DeVeaux Junior High School

Karen C. Rohrs
Hawkins Elementary School

Marie Sahloff
DeVeaux Junior High School

L. Michael Vince
McTigue Junior High School

Brenda D. Watkins
Old Orchard Elementary School

Oregon

Portland

Roberta Cohen
Catlen Gabel School

David Ellenberg
Catlen Gabel School

Sara Normington
Catlen Gabel School

Karen Scholte-Arce
Catlen Gabel School

West Linn

Marge Burack
Wood Middle School

Tracy Wygant
Athey Creek Middle School

Canby

Sandra Kralovec
Ackerman Middle School

Pennsylvania

Pittsburgh

Sheryl Adams
Reizenstein Middle School

Sue Barie
Frick International Studies Academy

Suzie Berry
Frick International Studies Academy

Richard Delgrosso
Frick International Studies Academy

Janet Falkowski
Frick International Studies Academy

Joanne George
Reizenstein Middle School

Harriet Hopper
Reizenstein Middle School

Chuck Jessen
Reizenstein Middle School

Ken Labuskes
Reizenstein Middle School

Barbara Lewis
Reizenstein Middle School

Sharon Mihalich
Reizenstein Middle School

Marianne O'Conner
Frick International Studies Academy

Mark Sammartino
Reizenstein Middle School

Washington

Seattle

Chris Johnson
University Preparatory Academy

Rick Purn
University Preparatory Academy

Contents

A basic knowledge of statistics and data analysis is a necessary life skill. Newspapers, television, billboards, and books bombard us with statistical data. Students need experience in both consuming and creating data that is represented in several different forms, including tables, graphs, and statistics.

Data About Us engages students in investigations about themselves. The unit introduces key concepts and processes in statistics and data analysis. As one of the first units in this curriculum, *Data About Us* is designed to help students learn to work together in groups. It will also help students realize what is expected in this curriculum, which focuses on building understanding through the investigation of problem situations.

The five investigations will help students learn more about ways to describe themselves and others. Each investigation includes one or more central problems, each of which has follow-up questions to help students focus on the embedded mathematics or on standard techniques for representing data. The investigations are framed to make it possible to encourage students' curiosity and to extend their considerations to other questions in addition to those posed.

We have assumed students have had some experience with data analysis and statistics in the elementary grades and that they have had opportunities to collect data themselves. The data sets presented in this unit have been selected based on their familiarity to students and the possibility that a class may want to collect their own data to analyze or to compare with the given data set. We encourage you to permit such exploration when feasible; in particular, if students have had little experience with data analysis and statistics, you may want to focus on the class's own data in several of the problems and ignore or delay consideration of the data sets provided with the materials.

Exploring statistics as a process of data investigation involves a set of four interrelated components (Graham, 1987):

- *Posing the question:* formulating the key question(s) to explore and deciding what data to collect to address the question(s)

- *Collecting the data:* deciding how to collect the data as well as actually collecting it

- *Analyzing the data:* organizing, representing, summarizing, and describing the data and looking for patterns in the data

- *Interpreting the results:* predicting, comparing, and identifying relationships and using the results from the analyses to make decisions about the original question(s)

This dynamic process often involves moving back and forth among the four interconnected components—for example, collecting the data and, after some analysis, deciding to refine the question and gather additional data. It may involve spending time working within a single component—for example, creating several different representations of the data, some in earlier stages of the process and others at a later time, before selecting the representation(s) to be used for final presentation of the data.

Reading Data Using Graphs

As a central component of data analysis, graphs deserve special attention. Curcio (1989) conducted a study of graph comprehension to assess the understanding of students in grades 4 and 7 of four traditional graphs: pictographs, bar graphs, circle or pie graphs, and line graphs. She identified three components to graph comprehension that are useful here:

- *Reading the data* involves "lifting" information from a graph to answer explicit questions. For example, How many students have 12 letters in their names?

- *Reading between the data* includes the interpretation and integration of information presented in a graph. For example, How many students have more than 12 letters in their names?

- *Reading beyond the data* involves extending, predicting, or inferring from data to answer implicit questions. For example, What is the typical number of letters in these students' names? If a new student joined our class, how many letters would you predict that student would have in his or her name?

Once students create their graphs, they use them in the interpretation phase of the data-investigation process. This is when they (and you) need to ask questions about the graphs. The first two categories of questions—*reading the data* and *reading between the data*—are basic to understanding graphs. However, it is *reading beyond the data* that helps students to develop higher-level thinking skills such as inference and justification.

Data *About Us* was created to help students

- Engage in the process of data investigation: posing questions, collecting data, analyzing data, and making interpretations to answer questions

- Represent data using line plots, bar graphs, stem-and-leaf plots, and coordinate graphs

- Explore concepts that relate to ways of describing data, such as the shape of a distribution, what's typical in the data, and measures of center (mode, median, mean, range or variability) in the data

- Develop a variety of strategies—such as using comparative representations and concepts related to describing the shape of the data—for comparing data sets

Investigation 1: Looking at Data

Students spend time talking about their names and how they were named. They consider the importance of lengths of names, which leads to an investigation of the distribution of numbers of letters in names. Students consider what's typical about a data set of lengths of names and then consider their class's data. They are introduced to or review the use of tables, line plots, and bar graphs to represent data; ways to describe the shape of a distribution; and the use of two measures of center (the mode and median) and a measure of spread (the range) that can be used to characterize a distribution.

Investigation 2: Kinds of Data About Us

Students are introduced to types of data, and they focus on categorical and numerical data. Two problems give students further experience with working with the two types of data. In the first problem, they write questions that will elicit responses that are either numerical or categorical. In the second problem, they consider two tables and graphs of data that relate to two questions, one numerical and one categorical.

Investigation 3: Using Graphs to Group Data

Data that are collected are often quite spread out or have great variability; a line plot or bar graph is not very useful for displaying such data. Students need strategies for grouping and displaying data in intervals. The stem-and-leaf plot (or stem plot) is a useful tool for grouping data in intervals of 10, and it helps students see patterns in the data. Students examine two given data sets—the first about time and distance required for students in a particular class to travel to school, and the second about how many times students in two different classes each jumped rope without stopping. These data provide the vehicle for introducing and exploring the stem plot.

Investigation 4: Measures About Us

Students use coordinate graphs to display pairs of data. They begin by collecting data about the lengths of their arm spans and their heights; using these data, they make a coordinate graph and sketch in the $y = x$ line so they can discuss people who are above, on, or below the line and what this means in terms of their own arm spans and heights. They then return to the travel time and distance data set and look at a coordinate graph that shows pairs of data (a student's travel time paired with distance traveled) in order to discuss relationships between these two data sets (that is, does it take someone who travels farther more time to get to school?).

Investigation 5: What Do We Mean by *Mean?*

This investigation focuses on developing the concept of *mean.* The "average" number of people in the families of students in your class provides the setting. The notion of "evening out" or "balancing" the distribution at a point (the mean) located on the horizontal axis is modeled by using interlocking cubes and stick-on notes. These models support the algorithm for finding the mean: adding up all the numbers and dividing by the total number of numbers.

Materials

For students

- Labsheets
- Index cards
- Interlocking cubes (10 each of 9 different colors per student)
- Stick-on notes
- Colored pens, pencils, or markers
- Scissors
- Large sheets of unlined paper
- Grid paper (provided as a blackline master)
- Yardsticks, meter sticks, or tape measures
- String (optional)

For the teacher

- Transparencies and tranparency markers (optional)
- Stick-on notes
- Colored stick-on dots (optional)
- Local street map (optional)
- Chart paper with a 1-inch grid (optional; available at an office-supply store)

Technology

The calculations in *Data About Us* involve only simple arithmetic, so nonscientific calculators are adequate. The unit may be pursued without access to software; however, if appropriate software is available, we encourage you to introduce your students to its use as part of this unit's activities. Use of such software may well improve students' understanding of the structure of graphs and will certainly promote their exploration of different graphs.

Resources

For students

The books below are good literary supplements to the topics in *Data About Us.*

Lee, Mary P. and Richard S. Lee. *Last Names First.* Philadelphia: Westminster Press, 1985.
Juster, Norton. *The Phantom Tollbooth.* New York: Alfred A. Knopf, 1961.
Chuks-Orji, Ogonna. *Names from Africa.* Chicago: Johnson Publishing Co., 1972.
Wolfman, Ira. *Do People Grow on Family Trees?* New York: Workman Publishing, 1991.

For teachers

Curcio, Fraces. *Developing Graph Comprehension.* Reston, Va.: National Council of Teachers of Mathematics, 1989.
Graham, Alan. *Statistical Investigations in the Secondary School.* Cambridge: Cambridge University Press, 1987.

Pacing Chart

This pacing chart gives estimates of the class time required for each investigation and assessment piece. Shaded rows indicate opportunities for assessment.

Investigations and Assessments	Class Time
1 Looking at Data	5 days
2 Types of Data	2 days
Check-Up 1	1/2 day
3 Using Graphs to Group Data	2 days
4 Coordinate Graphs	2 days
5 What Do We Mean by *Mean?*	5 days
Quiz	1 day
Check-Up 2	1 day
Self-Assessment	Take home
The Unit Project	Take home

Data About Us Vocabulary

The following words and concepts are introduced and used in *Data About Us.* Concepts in the left column are essential for student understanding of this and future units. The Descriptive Glossary/ Index describes these and other words used in *Data About Us.*

Essential

axis, axes

bar graph (bar chart)

categorical data

coordinate graph (scatter plot)

data

line plot

mean

median

mode

numerical data

outlier

range

scale

stem-and-leaf plot (stem plot)

survey

table

Nonessential

compare

measures of center (average)

predict, prediction

Embedded Assessment

Opportunities for informal assessment of student progress are embedded throughout *Data About Us* in the problems, the ACE questions, and Mathematical Reflections. Suggestions for observing as students discover and explore mathematical ideas, for probing to guide their progress in developing concepts and skills, and for questioning to determine their level of understanding can be found in the *Launch, Explore,* or *Summarize* sections of all investigation problems. Some examples:

- Investigation 1, Problem 1.2 *Launch* (page 21c) suggests ways to assess your students' understanding of bar graphs and line plots.
- Investigation 4, Problem 4.1 *Explore* (page 52c) suggests questions you might ask to help your students think about appropriate scales for the axes of their coordinate graphs.
- Investigation 5, Problem 5.3 *Summarize* (page 68h) provides help for you in guiding students to develop an algorithm for finding the mean of a set of data.

ACE Assignments

An ACE (Applications—Connections—Extensions) section appears at the end of each investigation. To help you assign ACE questions, a list of assignment choices is given in the margin next to the reduced student page for each problem. Each list indicates the ACE questions that students should be able to answer after they complete the problem.

Partner Quiz

One quiz, which may be given after Investigation 5, is provided with *Data About Us.* This quiz is designed to be completed by pairs of students with the opportunity for revision based on teacher feedback. The quiz involves class data. On the day before the quiz, ask your students to record the time they go to bed on a school night and pass this information in to you. Display this data (without student names) on the board as students take the quiz. You will find the quiz and its answer key in the Assessment Resources section. As an alternative to the quiz provided, you can construct your own quiz or quizzes by combining questions from the Question Bank, the quiz, and unassigned ACE questions.

Check-Ups

Two check-ups, which may be given after Investigations 2 and 5, are provided for use as quick quizzes or as warm-up activities. Check-ups are designed for students to complete individually. You will find the check-ups and their answer keys in the Assessment Resources section.

Question Bank

A Question Bank provides questions you can use for homework, reviews, or quizzes. You will find the Question Bank and its answer key in the Assessment Resources section.

Notebook/Journal

Students should have notebooks to record and organize their work. In the notebooks will be their journals along with sections for vocabulary, homework, and quizzes and check-ups. In their journals, students can take notes, solve investigation problems, record their mathematical reflections, and write down ideas for their unit projects. You should assess student journals for completeness rather than correctness; journals should be seen as "safe" places where students can try out their thinking. A Notebook Checklist and a Self-Assessment are provided in the Assessment Resources section. The Notebook Checklist helps students organize their notebooks. The Self-Assessment guides students as they review their notebooks to determine which ideas they have mastered and which ideas they still need to work on.

The Unit Project: Is Anyone Typical?

The final assessment for *Data About Us* is the Is Anyone Typical? Project. The project is introduced at the beginning of the unit, when students are asked to think about the different characteristics of middle-school students. As students complete each investigation, they are asked to write questions that they could use to collect data about the typical middle-school student, and to think about how what they are studying might help them organize and report data for their project. The project is formally assigned at the end of the unit. Students are asked to use all the concepts they have learned in *Data About Us* to create a survey to gather information about middle-school students.

Introducing Your Students to *Data About Us*

One way to introduce *Data About Us* is to have a class discussion about the "typical" middle-school student. Ask students what they think the word typical means. Then ask them what they think are some characteristics of a typical middle-school student. What is the typical height? The typical favorite musical group? The typical number of siblings?

Explain that, in this unit, students will gather and analyze data to try to find out some typical characteristics of their classmates. To do this, they first have to decide what information they want to know, and then write questions they could ask to gather this information. Have students suggest some things they would like to know about their classmates. Then, have them suggest questions they could ask to find out this information.

Discuss whether typical characteristics of your class would be typical of a class in another part of town, another state, or another country. For example, would the typical favorite food for students in your class be the same as the typical favorite food for a middle-school class in Japan?

Data About Us

What kinds of information would you like to know about students in your class?

What do we mean when we talk about a "typical" student at your grade level or in your school?

What things could you find out by comparing information about your class with information about other groups of students?

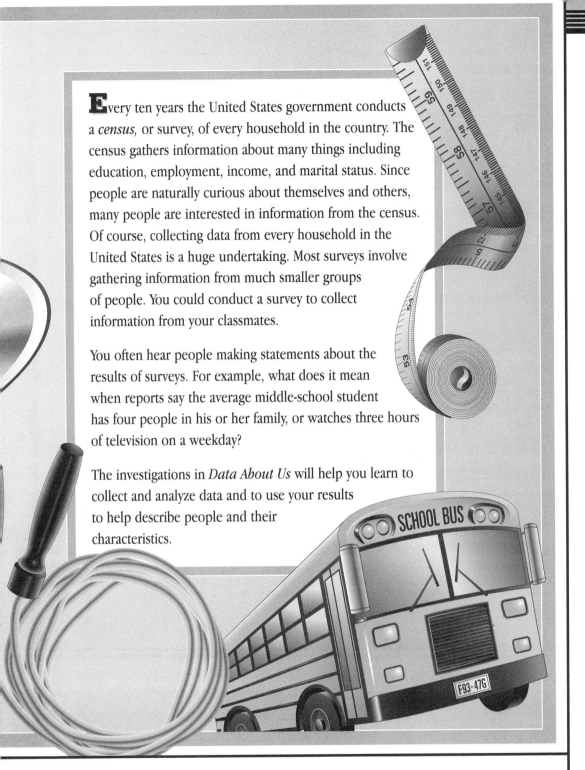

Every ten years the United States government conducts a *census,* or survey, of every household in the country. The census gathers information about many things including education, employment, income, and marital status. Since people are naturally curious about themselves and others, many people are interested in information from the census. Of course, collecting data from every household in the United States is a huge undertaking. Most surveys involve gathering information from much smaller groups of people. You could conduct a survey to collect information from your classmates.

You often hear people making statements about the results of surveys. For example, what does it mean when reports say the average middle-school student has four people in his or her family, or watches three hours of television on a weekday?

The investigations in *Data About Us* will help you learn to collect and analyze data and to use your results to help describe people and their characteristics.

After the discussion, refer students to page 5, where the unit project is introduced.

Mathematical Highlights

The Mathematical Highlights page provides information to students and to parents and other family members. It gives students a preview of the activities and problems in *Data About Us.* As they work through the unit, students can refer back to the Mathematical Highlights page to review what they have learned and to preview what is still to come. This page also tells parents and other family members what mathematical ideas and activities will be covered as the class works through *Data About Us.*

Mathematical Highlights

In *Data About Us,* you will learn how to collect, organize, analyze, and interpret data.

● Investigating data about the numbers of letters in the names of your classmates helps you learn how to organize data. When your data are organized, you can describe the "shape," find the *range* of values, and determine the *mode* (the value that occurs most frequently).

● Folding strips of grid paper containing data values leads you to discover the *median,* or middle value, in a data set. The median is an important kind of *average,* or measure of what is "typical" about a set of data.

● Making and interpreting line plots, bar graphs, stem-and-leaf plots, and coordinate graphs lets you see the kinds of information each graph displays and helps you choose the most appropriate graph for a given set of data.

● Data about pets illustrate the two types of data—*categorical data* and *numerical data*.

● "Evening out" cube towers leads you to discover the *mean,* another important kind of average.

● Adding and removing values from data sets and recomputing the averages shows you how the mean and median are affected by *outliers* and changes in data.

● Performing your own statistical investigation to try to discover what is typical about your classmates ties together what you learn in this unit and helps you see the power of statistics.

The Unit Project

Is Anyone Typical?

What are the characteristics of a typical middle-school student? Who would be interested in knowing these characteristics? Does a typical middle-school student really exist? As you proceed through this unit, you will identify some "typical" facts about your classmates, such as these:

- The typical number of letters in a student's full name
- The typical number of people in a student's household
- The typical height of a student

When you have completed the investigations in *Data About Us,* you will carry out a statistical investigation to answer this question: What are some of the characteristics of a typical middle-school student? These characteristics may include

- physical characteristics (for example, age, height, or eye color)
- family and home characteristics (for example, number of brothers and sisters or number of television sets)
- miscellaneous behaviors (for example, hobbies or number of hours spent watching television)
- preferences, opinions, or attitudes (for example, favorite musical group, or opinions about who should be elected class president)

Keep in mind that a statistical investigation involves posing questions, collecting data, analyzing data, and interpreting the results of the analysis. As you work through each investigation, think about how you might use what you are learning to help you with your project.

Introducing the Unit Project

The final assessment for *Data About Us,* the Is Anyone Typical? project, is a research project in which students have the opportunity to develop, conduct, evaluate, and report the results of a survey. The project is introduced on this page and formally assigned at the end of the unit. To introduce the project, students are asked to think about characteristics of middle-school students. Throughout the unit, students are reminded to use the concepts they are learning to write questions they might ask in their surveys and to think about how they will analyze and interpret the data they collect. Some teachers have found it useful to have students designate two or three "Is Anyone Typical?" pages in their journals to record this information. At the end of the unit, students are asked to apply the concepts they have learned in *Data About Us* to develop a survey and then to analyze and display the data they collect.

See pages 68 and 86 for information about assigning the project. To help you assess the projects, see page 88. Here you will find a possible scoring rubric and samples of student projects.

The Investigations

The teaching materials for each investigation consist of three parts: an overview, the student pages with teaching outlines, and the detailed notes for teaching the investigation.

The overview of each investigation includes brief descriptions of the problems, the mathematical and problem-solving goals of the investigation, and a list of necessary materials.

Essential information for teaching the investigation is provided in the margins around the student pages. The "At a Glance" overviews are brief outlines of the Launch, Explore, and Summarize phases of each problem for reference as you work with the class. To help you assign homework, a list of "Assignment Choices" is provided next to each problem. Wherever space permits, answers to problems, follow-ups, ACE questions, and Mathematical Reflections appear next to appropriate student pages.

The Teaching the Investigation section follows the student pages and is the heart of the Connected Mathematics curriculum. This section describes in detail the Launch, Explore, and Summarize phases of each problem. It includes all the information needed for teaching, along with suggestions for what you might say at key points in the teaching. Use this section to prepare lessons and as a guide for teaching an investigation.

Assessment Resources

The Assessment Resources section contains blackline masters and answer keys for the quiz, check-ups, and the Question Bank. It also provides guidelines for assessing the unit project and other important student work. Samples of student work, along with the teacher's comments about how each sample was assessed, will help you to evaluate your students' efforts. Blackline masters for the Notebook Checklist and the Self-Assessment support student self-evaluation, an important aspect of assessment in the Connected Mathematics curriculum.

Blackline Masters

The Blackline Masters section includes masters for all labsheets and transparencies. A blackline master of grid paper is also provided.

Descriptive Glossary/Index

The Descriptive Glossary/Index provides descriptions and examples of the key concepts in *Data About Us*. These descriptions are not intended to be formal definitions, but are meant to give you an idea of how students might make sense of these important concepts. The page number references indicate where each concept is first introduced.

Looking at Data

Mathematical and Problem-Solving Goals

- **To use tables, line plots, and bar graphs to display data**

- **To use measures of center (mode and median) and measures of spread (range and intervals within the range) to describe what is typical about data**

- **To describe the shape of the data**

- **To experiment with how the median, as a measure of center, responds to changes in the number and magnitude of data values**

This first investigation develops some introductory statistical techniques that will be used throughout *Data About Us.*

Although in many of the problems data are provided, we encourage you to have your class collect their own data for some of the problems. Keep in mind that collecting data is time-consuming, so carefully choose the problems for which you will have students generate data.

In Problem 1.1, Organizing Your Data, students collect and organize data about the numbers of letters in their names. This open-ended problem will help you to assess the techniques your students have developed to organize, summarize, and display data.

In Problem 1.2, Interpreting Graphs, students compare a line plot and a bar graph representing the same data. This problem helps students to review (or to learn) the structure and use of these types of graphs. In Problem 1.3, Identifying the Mode and Range, students create data distributions with a given range and mode. This helps them to recognize the parts that the mode and range play in providing an overall description of a data set. In Problem 1.4, Identifying the Median, students fold paper strips in half to model finding the median. In Problem 1.5, Experimenting with the Median, students investigate how the median of a data set changes as values are added or removed.

Student Pages	6–21
Teaching the Investigation	21a–21k

Materials

For students

- Index cards (20 per group)
- Grid paper
- Scissors
- Large sheets of unlined paper (optional)
- Stick-on notes or interlocking cubes (optional)

For the teacher

- Transparencies 1.1, 1.2, 1.3, 1.4A, 1.4B, and 1.5 (optional)
- Stick-on notes
- Chart paper with a 1-inch grid (optional; available at an office-supply store)

Launch

- Engage students in a discussion of how they acquired their names and situations in which name length is important.

- Pose the question of how to collect data on the students' name lengths and how to define exactly what data need to be collected.

Explore

- Circulate while students collect data in groups or as a class, and help them to organize the information and to design and label tables and graphs.

- Take note of groups with particularly good methods of organizing the data.

Summarize

- Allow a few groups to share their results.

- Lead a discussion about whether there is a *typical* name length.

INVESTIGATION 1

Looking at Data

The problems in this investigation involve people's names. Names are filled with symbolism and history. Because family traditions are often involved when a child is named, a person's name may reveal information about his or her ancestors.

Many people have interesting stories about how they were named. Here is one student's story of how her name was chosen: "I'm a twin, and my mom and dad didn't know they were going to have twins. My sister was born first, and she was named Susan. I was a surprise. My mom named me after the woman in the next hospital bed, whose name was Barbara."

Compare stories with your classmates about how you, or someone you know, were named.

1.1 Organizing Your Data

Most parents spend little time worrying about the number of letters in the names they choose for their children. Yet there are times that name length matters. For example, there is sometimes a limit to the number of letters that will fit on a friendship bracelet or a library card.

Did you know?

The longest name appearing on a birth certificate is Rhoshandiatellyneshiaunneveshenk Koyaanfsquatsiuty Williams.

Shortly after Rhoshandiatellyneshiaunneveshenk was born, her father filed an amendment that expanded her first name to 1019 letters and her middle name to 36 letters. Can you think of a good nickname for her?

Source: *Guinness Book of World Records*

Assignment Choices

None

Answers to Problem 1.1

Answers will vary. See the "Summarize" section on page 21b.

Answers to Problem 1.1 Follow-Up

Answers will vary. See the "Summarize" section on page 21b.

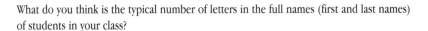

What do you think is the typical number of letters in the full names (first and last names) of students in your class?

Problem 1.1

Gather data about the total number of letters in the first and last names of students in your class.

A. Find a way to organize the data so you can determine the typical name length.

B. Write some statements about your class data. Note any patterns you see.

C. What would you say is the typical name length for a student in your class?

D. If a new student joined your class today, what would you predict about the length of that student's name?

Launch

- Discuss Problem 1.2, and ask the class to consider how the line plot and the bar graph were made.

- Ask questions to help students understand the two representations.

▓ **Problem 1.1 Follow-Up**

Do you think the length of your name is typical for a student in your class? Explain why or why not.

1.2 Interpreting Graphs

A group of students in Ms. Jeckle's class made a line plot to display their class's name-length data.

Name Lengths of Ms. Jeckle's Students

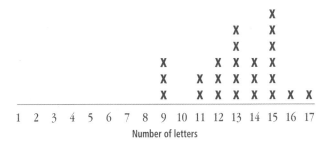

Number of letters

Explore

- Have students work on the problem at home or in class.

- If students explore the problem in class, ask questions to keep them focused on the problem.

Summarize

- As a class, discuss the answers to the problem and the follow-up.

- Help clarify for students the specific features of line plots and bar graphs.

Answers to Problem 1.2

A. Possible answers: There is one peak at a name length of 15 letters. The length of the names ranges from 9 letters to 17 letters. Only two of the names are longer than 15 letters. Most of the name lengths cluster in the interval of 12–15 letters.

B. Possible answers: Both graphs have titles and labeled axes. They have similar horizontal scales, which show the possible name lengths. The line plot does not have a vertical scale; the frequency of a particular name length is indicated by the number of Xs above that number. The bar graph has a vertical scale, which is needed to determine the frequencies based on the height of each bar.

C. Answers will vary.

Another group displayed the same data in a bar graph.

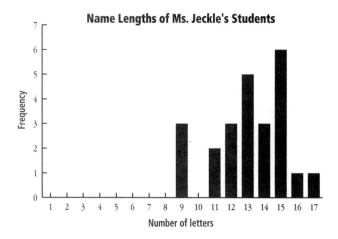

Name Lengths of Ms. Jeckle's Students

Number of letters

Problem 1.2

Examine the line plot and the bar graph.

A. Write some statements about the name lengths for students in Ms. Jeckle's class. Describe any interesting patterns you see in the data.

B. In what ways are the two graphs alike? In what ways are they different?

C. How does the data from Ms. Jeckle's class compare with the data from your class?

■ **Problem 1.2 Follow-Up**

1. How can you use each graph to determine the total number of letters in all the names?

2. Fahimeh Ghomizadeh said, "I have the most letters in my name, but the bar that indicates the number of letters in my name is one of the shortest. Why?" How would you answer this question?

3. Suppose a new student named Nicole Martin joined Ms. Jeckle's class. How could you change the graphs to include data for Nicole?

Answers to Problem 1.2 Follow-Up

1. For the bar graph, you could multiply the height of each bar—which represents a number of people—by the name length each bar represents. For example, three people have nine letters in their name, so the bar over the 9 accounts for 9 × 3, or 27, letters. Add the results for each bar to find the total number of letters. For the line plot, you could multiply the number of Xs over each name length by that name length and add the results.

2. The height of a bar does not represent a name length; it represents the number of people with a particular name length. Since Fahimeh is the only student who has a name with 17 letters, the bar over the 17 has a height of 1.

3. You would need to add another X above the 12 on the line plot. You would need to lengthen the bar above the 12 on the bar graph to a height of 4.

Did you know?

In parts of Africa, a child's name is often very meaningful. Names such as Sekelaga, which means "rejoice," and Tusajigwe, which means "we are blessed," reflect the happiness the family felt at the child's birth. Names such as Chukwueneka, which means "God has dealt kindly with us," demonstrate the family's religious feelings. Names such as Mvula, meaning "rain," reflect events that happened at the time the child was born.

1.3 Identifying the Mode and Range

One way to describe what is typical, or average, about a set of data is to give the value that occurs most frequently. For example, in the data for Ms. Jeckle's class, the name length 15 occurs most frequently. Six students have 15 letters in their names. Notice that 15 has the highest stack of X's in the line plot and the tallest bar in the bar graph. We call the value that occurs most frequently the **mode** of the data set.

When describing a data set, it is also helpful to give the lowest value and the highest value that occur. The spread of data values from the lowest value to the highest value is called the **range** of the data. In Ms. Jeckle's class, the range of name lengths is from 9 letters to 17 letters.

Think about this!

What are the mode and range of the name-length data for your class? How do the mode and range for your class compare with the mode and range for Ms. Jeckle's class?

At a Glance

Launch

- As a class, discuss the ideas of *mode* and *range*.

- Read and clarify the details of Problem 1.3.

Explore

- Have groups explore the problem and construct line plots to display the data for the class to view.

Summarize

- As a class, compare the various line plots students created.

- Have students consider how the constraints on the problem affected the possible shapes of the data.

Tips for the Linguistically Diverse Classroom

Rebus Scenario The Rebus Scenario technique is described in detail in *Getting to Know CMP*. This technique involves sketching rebuses on the chalkboard that correspond to key words in the story or information you present orally. Example: some key words for which you may need to draw rebuses while discussing the Did you know? feature—*Africa* (an outline of the continent), *Sekelaga/rejoice* (face with a huge smile), *Mvula/rain* (rain cloud with rain).

Assignment Choices

ACE questions 3–5, 7, 11–14, and unassigned choices from earlier problems

Problem 1.3

There are 15 students in a class. The mode of the name lengths for the class is 12 letters, and the range is from 8 letters to 16 letters.

A. Determine a set of name lengths that has this range and mode.

B. Make a line plot to display your data.

C. Use your line plot to help you describe the shape of your data. For example, your data may be bell-shaped, spread out in two or more clusters, or grouped together at one end of the graph.

■ **Problem 1.3 Follow-Up**

Compare your graph with the graphs some of your classmates drew. How are the graphs alike? How are they different?

Did you know?

Here are some interesting facts about family names.

• It wasn't until the 1800s that countries in eastern Europe and Scandinavia insisted that people adopt permanent family names.

• The most common family name in the world is Chang. An estimated 100 million Chinese people have this name.

• The most common name in the United States, Canada, and the United Kingdom is Smith. There are approximately 2.3 million Smiths in the United States alone.

• There are over 1.6 million different family names in the United States.

Answers to Problem 1.3

A. Answers will vary. A correct answer will have 15 values between 8 and 16, with the value 12 occurring most frequently.

B. See page 21i.

C. Answers will vary.

Answer to Problem 1.3 Follow-Up

Answers will vary. All the graphs should have a total of 15 Xs, a mode of 12, and a range of from 8 letters to 16 letters.

1.4 Identifying the Median

You have learned that one way to describe what is typical about a set of data is to give the value that occurs most frequently (the mode). Another way to describe what is typical is to give the middle value of the data set.

The table and line plot below show name-length data for a middle-school class in Michigan. Notice that this data has two modes, 11 and 12. The range of the data is from 8 letters to 19 letters.

Name	Letters
Jeffrey Piersonjones	19
Thomas Petes	11
Clarence Surless	15
Michelle Hughes	14
Shoshana White	13
Deborah Locke	12
Terry Van Bourgondien	19
Maxi Swanson	11
Tonya Stewart	12
Jorge Bastante	13
Richard Mudd	11
Joachim Caruso	13
Roberta Northcott	16
Tony Tung	8
Joshua Klein	11
Janice Vick	10
Bobby King	9
Jacquelyn McCallum	17
Kathleen Boylan	14
Peter Juliano	12
Linora Haynes	12

Class Name Lengths

```
              X  X
              X  X  X
              X  X  X  X                 X
     X  X  X  X  X  X  X  X  X  X  X      X
     7  8  9  10 11 12 13 14 15 16 17 18 19 20
                 Number of letters
```

Identifying the Median

At a Glance

Launch

- Discuss the idea of *averages* or *measures of center,* including the median and the mode.

- As a class, inspect the data for Problem 1.4, and arrange the values in ascending order.

Explore

- Have groups explore the problem.

- If any students finish early, have them explore the question posed in the "Think about this!" box.

Summarize

- As a class, explore the median on the strips of 21 and 22 values, and consider the median and the mode(s) of the two data sets.

- Talk about possible ways of finding the median for a large set of data.

- Have students work on the follow-up at home or in class, and discuss the answers.

Assignment Choices

ACE questions 2, 6, 8, 9, 10, and unassigned choices from earlier problems

Problem 1.4

Cut a strip of 21 squares from a sheet of grid paper. Write the Michigan class's name lengths in order from smallest to largest on the grid paper as shown here.

| 8 | 9 | 10 | 11 | 11 | 11 | 11 | 12 | 12 | 12 | 12 | 13 | 13 | 13 | 14 | 14 | 15 | 16 | 17 | 19 | 19 |

Now, put the ends together and fold the strip in half.

A. Where does the crease land? How many numbers are to the left of the crease? How many numbers are to the right of the crease?

Suppose a new student, Suzanne Mannerstrale, joins the Michigan class. The class now has 22 students. On a strip of 22 squares, list the name lengths, including Suzanne's, in order from smallest to largest. Fold this strip in half.

B. Where is the crease? How many numbers are to the left of the crease? How many numbers are to the right of the crease?

■ Problem 1.4 Follow-Up

The first strip of paper had 21 data values. When you folded the strip, the crease was on the number 12. There were ten values to the left of 12 and ten values to the right of 12. We say that 12 is the *median* of the data set. The **median** of a data set is the value that divides the data in half—half of the values are below the median, and half the values are above the median.

The second strip you made had 22 values. When you folded this strip, the crease landed between 12 and 13. There were eleven values to the left of the crease and eleven values to the right of the crease. When a data set has an even number of values, the median is the value halfway between the two middle values. For this data set, the median is $12\frac{1}{2}$, the number halfway between 12 and 13.

Giving the median of a set of data is one way to describe what is typical about the data. Like the mode, the median is a type of *average*. The median and the mode are sometimes referred to as *measures of center*. You can see that this is a very appropriate description for the median, since it *is* the center of the data.

1. Find the median name length for your class.
2. Use the median, mode, and range to describe what is typical about your class's data.
3. Suppose a student named Chamique Holdsclaw joins your class. Add Chamique's name to your class data, and find the new median. How does the median change?

Answers to Problem 1.4

A. The crease lands on 12. There are 10 values to the right of the crease and 10 values to the left of the crease.

B. The crease lands between 12 and 13. There are 11 values to the right of the crease and 11 values to the left of the crease.

Answers to Problem 1.4 Follow-Up

Answers will vary.

Think about this!

For a very large data set, the method of folding a strip of data values would not be very efficient. Try to develop a different strategy for finding the median of a large data set.

1.5 Experimenting with the Median

What happens to the median when you add values to or remove values from a set of data? Does adding a value that is much larger or much smaller than the rest of the data values have a greater effect on the median than adding a value that is closer to the other values?

Write each of the names listed below on an index card. On the back of each card, write the number of letters in the name.

Name	Letters
Thomas Petes	11
Michelle Hughes	14
Shoshana White	13
Deborah Locke	12
Tonya Stewart	12
Richard Mudd	11
Tony Tung	8
Janice Vick	10
Bobby King	9
Kathleen Boylan	14

Richard
Mudd

front

11
letters

back

Order the cards from shortest name to longest name, and find the median of the data.

At a Glance

Launch

■ Have groups prepare their index cards, arrange them in ascending order, and determine the median of the values.

■ As a class, explore how to remove two cards without altering the median.

Explore

■ As groups explore the problem, help them to focus on how changes in the data affect the median.

Summarize

■ Have groups give examples of how to meet each set of criteria.

■ Help students to understand the stability of the median.

Answers to Problem 1.5

A. Possible answers: Remove the lowest value, 8, and the highest value, 14; the median remains $11\frac{1}{2}$. Remove the two middle values; the median remains $11\frac{1}{2}$.

B. Possible answers: Remove 8 and 9; the median increases to 12. Remove 10 and 11; the median increases to 12.

C. Possible answers: Remove both 12s; the median decreases to 11. Remove both 14s; the median decreases to 11.

D. Possible answers: Add 16 and 17; the median increases to 12. Add 900 and 1000; the median increases to 12.

E. Possible answers: Add 4 and 5; the median decreases to 11. Add 11 and 11; the median decreases to 11.

(Continued on next page.)

Assignment Choices

Any unassigned choices from earlier problems

Problem 1.5

Experiment with your cards to see if you can perform each task described below. Keep a record of the things you try and the discoveries you make.

A. Remove two names without changing the median.

B. Remove two names so the median increases.

C. Remove two names so the median decreases.

D. Add two new names so the median increases.

E. Add two new names so the median decreases.

F. Add two new names without changing the median.

■ **Problem 1.5 Follow-Up**

1. If a name with 16 letters were added to the data, what would the new median be?

2. If a name with 1019 letters were added to the data, what would the new median be?

Did you know?

Names from many parts of the world have special origins. European family names were often based on the father's first name. For example, Ian Robertson was the son of Robert, and Janos Ivanovich was the son (vich) of Ivan. Sometimes, the father's first name was used "as is" or with an "s" added to the end. For example, John Peters was the son of Peter, and Henry James was the son of James. Surnames were also created from words that told where a person lived, what a person did, or described personal characteristics. This resulted in names like William Hill, Geoffrey Marsh, Sean Forest, Gilbert Baker, James Tailor, and Kyle Butcher.

Surnames in China and Vietnam often have a long history and are almost always one-syllable words related to names of ruling families. Chang—a name mentioned earlier—is one such example.

Jewish names are sometimes made up of abbreviations that combine a number of words: Katz comes from *kohen tzedek* (righteous priest), and Schatz from *shalian tzibur* (representative of the congregation).

You can read more about names in books such as *Names from Africa* by O. Chuks-orji and *Do People Grow on Family Trees?* by Ira Wolfman.

F. Possible answers: Add 11 and 12; the median remains $11\frac{1}{2}$. Add 1 and 2000; the median remains $11\frac{1}{2}$.

Answers to Problem 1.5 Follow-Up

1. The median would increase to 12.

2. The median would increase to 12.

Applications • Connections • Extensions

As you work on these ACE questions, use your calculator whenever you need it.

Applications

For 1 and 2, use the names listed below.

Ben Carter
Ava Baker
Sarah Edwards
Juan Norinda
Ron Weaver
Bryan Wong
Toby Vanhook
Katrina Roberson
Rosita Ramirez
Kimberly Pace
Paula Wheeler
Darnell Cox
Jessica Otto
Erin Froyeh
Corey Buysse
Tijuana Degraffenreid

1. Make a table showing the length of each name. Then make both a line plot and a bar graph of the name lengths.

2. What is the typical name length for this class of students? Use the mode, median, and range to help you answer this question.

Applications

1. See page 21j.

2. The median name length is 11 letters, and the range of the data is 8–20 letters. A name length of 20 letters is somewhat unusual. The typical number of letters is clustered around the median in an interval of 8–13 letters or 9–12 letters. The mode is the same as the median in this example, although there are two other lengths that occur almost as frequently as the mode—9 letters and 12 letters.

3. 15; the mode

4. 28; The bar for each number represents the number of students with that name length, so adding the bar heights (1 + 2 + 4 + 3 + 4 + 7 + 3 + 2 + 2) gives the total number of students.

5. 10 letters to 19 letters

6. $14\frac{1}{2}$; There are 28 name lengths, so the median is between the fourteenth and fifteenth numbers, which are 14 and 15.

7. See below right.

8–10. See page 21k.

In 3–6, use the bar graph below.

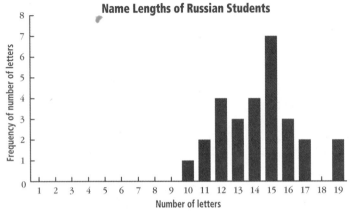

3. Which value (name length) occurs most frequently? What do we call this value?

4. How many students are in this class? Explain how you got your answer.

5. What is the range of name lengths for this class?

6. What is the median name length? Explain how you got your answer.

For 7–10, make a line plot or bar graph of a set of data that fits the description.

7. 24 names, with a range from 8 letters to 20 letters

8. 7 names, with a median length of 14 letters

9. 13 names, with a range from 8 letters to 17 letters and a median of 13 letters

10. 16 names, with a median of $14\frac{1}{2}$ letters and a range from 11 letters to 20 letters

Connections

In 11–14, use the bar graphs on page 17, which show information about a class of middle-school students.

7. Possible answer:

Name Lengths

```
            X  X
            X  X  X              X
            X  X  X  X  X        X              X
   X        X  X  X  X  X  X  X  X        X
   ─────────────────────────────────────────────
   7  8  9  10 11 12 13 14 15 16 17 18 19 20 21
              Number of letters
```

Graph A

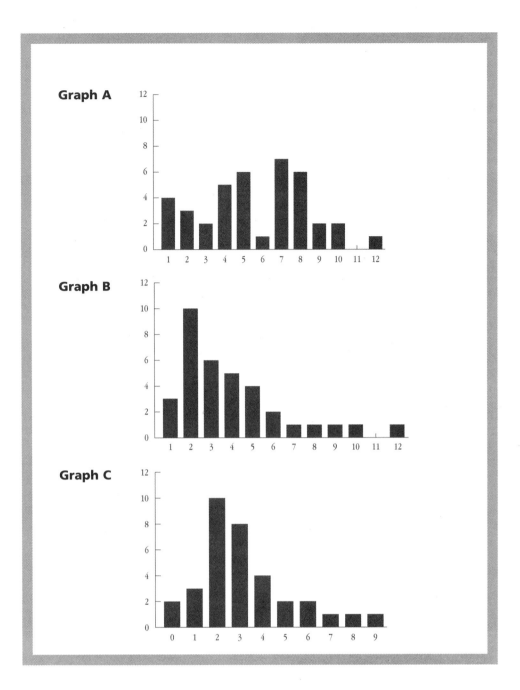

Graph B

Graph C

Connections

11. Possible answer: "Graph B"; "Graph C" has values that are 0; since the students are children, their families could not have 0 children. "Graph A" shows many values of 5, 7, and 8. It isn't likely that there will be a lot of families with these numbers of children.

12. "Graph A"; Students may also say that "Graph B" is correct since it is also labeled 1 through 12 on the horizontal. However, the distribution in "Graph B" is very unlikely for this data. "Graph C" cannot represent birth months, since it has a value at 0.

13. Any of the graphs could show numbers of pizza toppings. Many students argue that "Graph C" is correct because it could not show children in students' families or birth months. Others argue for "Graph C" because most of their friends like two or three toppings.

14. Possible answers:

Graph A—title: "Birth Months of Students"; horizontal axis label: "Birth month"; vertical axis label: "Frequency"

Graph B—title: "Number of Children in Students' Families"; horizontal axis label: "Number of children"; vertical axis label: "Frequency"

Graph C—title: "Number of Pizza Toppings"; horizontal axis label: "Number of toppings"; vertical axis label: "Frequency"

11. Which graph might show the number of children in the students' families? Explain your choice.

12. Which graph might show the birth months of the students? (Hint: Months are often written using numbers instead of names. For example, 1 means January, 2 means February, and 3 means March.) Explain your choice.

13. Which graph might show the number of toppings students like on their pizzas? Explain your choice.

14. Give a possible title, a label for the vertical axis, and a label for the horizontal axis for each graph based on your answers to 11–13.

Extensions

In 15–21, use the table and bar graphs below. A greeting card store sells stickers and street signs with first names on them. The store ordered 12 packages of stickers and 12 street signs for each name. The four bar graphs show the numbers of sticker packages and street signs that remain for the names that begin with the letter A.

Name	Stickers remaining	Street signs remaining
Aaron	1	9
Adam	2	7
Alice	7	4
Allison	2	3
Amanda	0	11
Amber	2	3
Amy	3	3
Andrea	2	4
Andrew	8	6
Andy	3	5
Angela	8	4
Ann	10	7

Graph A:
Stickers Remaining

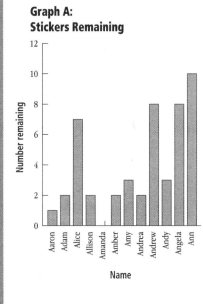

Graph B:
Street Signs Remaining

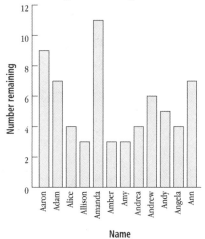

Graph C:
Stickers and Street Signs Remaining

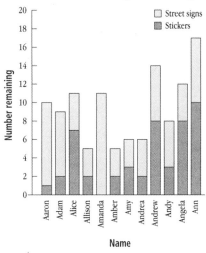

Graph D:
Stickers and Street Signs Remaining

Extensions

15. The bar height, 7, represents the number of stickers left. Since there were 12 packages to begin with, 5 have been sold.

16. The bar height, 4, represents the number of street signs left. Since there were 12 signs to begin with, 8 have been sold.

17. The bar graphs show that the number of stickers remaining is less than the number of street signs remaining. Students may want to debate this because of the "peaks" in the data. You will want to remind them that the bar graphs show the number remaining, not the number sold.

18. The store has taken in $144 on the sale of 96 name stickers.

19. The most stickers, 12, have been sold for Amanda. The fewest stickers, 2, have been sold for Ann.

20. For Amy, the bars for stickers and signs are the same height. This shows that equal numbers of sticker packages and signs have been sold for Amy.

21. The stacked bars show the data for stickers and street signs together. For example, Amanda has 0 stickers and 11 street signs left, while Alice has 7 stickers and 4 street signs remaining. However, the stacked bars for these names are the same height. Allison and Amber are the most popular names, while Ann is the least popular.

15. In "Graph A" locate the bar that shows the number of stickers left for the name Alice. Explain how you can determine how many stickers are left by reading the graph. Explain how you can determine how many stickers have been sold.

16. In "Graph B" locate the bar that shows the number of street signs left for the name Alice. Explain how you can determine how many street signs are left by reading the graph. Explain how you can determine how many street signs have been sold.

17. Are the stickers more popular than the street signs? Explain your answer.

18. If each package of stickers costs $1.50, how much money has the store made from selling name stickers for names beginning with A?

19. For which name have the most stickers been sold? For which name have the fewest stickers been sold?

20. "Graph C" is a *double bar graph*. Use this graph to determine the name(s) for which the number of street signs sold and the number of sticker packages sold are the same.

21. "Graph D" is a *stacked bar graph*. Use this graph to determine whether some names are more popular than others. Justify your answer.

Mathematical Reflections

In this investigation, you learned some ways to describe what is typical about a set of data. The following questions will help you summarize what you have learned:

① How are a table of data, a line plot, and a bar graph alike? How are they different?

② What does the mode tell you about a set of data?

③ What does the median tell you about a set of data?

④ Can the mode and the median for a data set be the same? Can they be different? Explain your answers.

⑤ Why is it helpful to give the range when you describe a set of data?

⑥ What does it mean to describe the shape of the data?

⑦ How can you describe what is typical about a set of data?

Think about your answers to these questions, discuss your ideas with other students and your teacher, and then write a summary of your findings in your journal.

At the end of this unit, you will be developing a survey to gather information about middle-school students. Think of a question or two you might ask. How would you display the information you might gather about each of these questions? Write your thoughts in your journal.

Tips for the Linguistically Diverse Classroom

Diagram Code The Diagram Code technique is described in detail in *Getting to Know CMP*. Students use a minimal number of words and drawings or diagrams to respond to questions that require writing. Example: Question 3—A student might answer this question by drawing a simple line plot with a thick line dividing the data in half with the words $\frac{1}{2}$ *data* on each side of the line.

Possible Answers

1. A table of data, a line plot, and a bar graph are all tools for organizing and visualizing data. All three indicate the possible values of the item being measured (for example, 10 letters, 11 letters). A line plot and a bar graph indicate the number of times each value occurs.

A line plot has a horizontal axis that shows the possible values, and marks above the numbers indi- ⌐ cating the number of times each value occurs.

Like the line plot, a bar graph has a horizontal axis showing the possible values. Instead of using marks, the number of times a value occurs is indicated by the height of the bar over the value. A vertical axis indicates the frequency, corresponding to the height of each bar.

2. The mode is the value in a data set that occurs most frequently. There may be more than one mode, and a mode may occur at any location in the data.

3. The median is the value that divides an ordered set of data in half; half the data are below the median, and half the data are above the median. The median is not easily affected by the addition of very high or very low values.

4–7. See page 21k.

TEACHING THE INVESTIGATION

1.1 • Organizing Your Data

Launch

Students begin by considering how they were named. Help to engage them in a short discussion of names.

> Do you know anything interesting about how you were named or about the history behind your family's name?

Give students some time to discuss their stories.

Problem 1.1 asks students to gather data about their names and to organize their data so they can see patterns and determine the typical name length for students in the class.

> You probably don't spend much time thinking about the number of letters in your name, but there are times when the length of your name matters. Can you think of a situation in which it would matter?

Students may have several ideas. Many will be familiar with scannable answer sheets on which they record their names by filling in boxes or bubbles, and some may have found that there are not enough spaces to record their full names.

> What do you think is the typical number of letters in the full name of a student in this class? Before we can answer this question, we need to decide as a class what we will consider to be a full name.

Part of statistical investigation involves determining exactly what data need to be collected, so it is important that the class reaches a consensus about what constitutes a full name. Students will need to discuss exactly what letters they will count. Do they use nicknames, or full first names? Do they use middle initials, or full middle names? Does the hyphen in a name such as Clarke-Peterson or Mai-Lin count as a letter?

After the class has defined the data, have students work in small groups to collect the data, or gather the data as a class. Here are some suggestions of how you might gather the data.

- Each student can use cubes to make a stack with the same number of cubes as letters in his or her name. The stacks can then be placed where the whole class can see them.

- Students can take turns calling out the number of letters in their names while you record each number on the board.

If you collect the data as a class by recording the counts on the board or at the overhead, be sure to collect it in a way that does not automatically organize it. Once the data are collected, have students work in groups to devise a plan for organizing the data.

> How do you think we should organize this information? Suppose you wanted to tell another class about the name lengths of students in our class. It would be helpful to first organize the data so you can see patterns and determine what a typical name length is.

Explore

As groups are considering the problem, you may need to ask questions to keep students working productively. If students are stuck, here are suggestions you could make to get them started:

- Arrange the name lengths from smallest to largest.
- Create a table showing each name and the number of letters it contains.

Most groups will simply write the name lengths in order from smallest to largest. Ask them questions to help them think of other ways to organize the data.

> Can you tell from your list where the clusters of name lengths are?
>
> Can you tell what name length occurs most frequently?
>
> What name lengths do not occur in our class?
>
> What other ways might you organize the data so you can answer questions like these?

As you circulate, take note of which groups may be able to help the class to consider better ways to display the data. When most groups feel they have something to share, move on to the summary.

Summarize

Give some of the groups time to show and explain their work, including the groups you noticed that had found good ways to display the data.

Emphasize that students should label their tables and graphs so others can "read" the representations. Tables and graphs should have titles, and table columns and graph axes should be labeled. Discuss titles for tables and graphs and labels for columns and axes.

When students have displayed their data, they can consider ways to describe the *shape of the data*—what the distribution looks like.

> If we wanted to describe our data to another class, what might we say about how it looks?

Students should note overall patterns in the data. The data may be spread out, or may fall into one or more clusters. Students should also consider whether there is a value that is *typical*, or representative, of the data set. Some students may indicate the mode as a typical value. Noting an interval of values (for example, that most names have lengths of 11–13 letters) is also an acceptable way of describing what is typical. After discussing the questions in the follow-up, you might ask other hypothetical questions:

> If we did this activity using names of students in another class, what would you expect to find?

Encourage students to relate this prediction to the name length they think is typical.

> If a new student joined our class today, what would you predict to be the number of letters in his or her name?

Students may indicate that they expect a new student's name length to fall within what they consider to be the typical range of number of letters. They may also indicate that they expect it to fall in an area of the display with fewer responses because they anticipate the "holes" will be filled in eventually.

1.2 • **Interpreting Graphs**

Launch

This problem reviews or teaches students about line plots and bar graphs and relates the structures of these two types of graphs. Research has shown that students do not automatically "translate across representations"—they may not initially view the line plot and the bar graph as comparable representations and may not recognize when a line plot and a bar graph display identical information. This problem provides an opportunity for you to assess and reinforce students' understanding.

Show Transparency 1.2 (or refer students to pages 7 and 8 in their books), and then read Problem 1.2 aloud. Ask the class to explain how they think the line plot and the bar graph were made.

What do the Xs on the line plot represent?

What do the bars on the bar graph represent?

What do each of the axes represent?

How does the bar chart show the X that is above 17 in the line plot?

When you feel most students understand the two representations, let them work on the problem.

Explore

You may want to have your students work on Problem 1.2 at home. If you do the problem in class, ask questions as you circulate to help keep students focused on the problem.

Summarize

Begin the summary by discussing the answers to the problem.

Spend some time discussing the questions in Problem 1.2 Follow-Up. These questions will help students focus on important features of the graphs. They may have greater difficulty than you might imagine with these kinds of questions.

1.3 • Identifying the Mode and Range

Launch

Use the data set from Problem 1.2 to explain the words *mode* and *range.* You can display Transparency 1.2 on the overhead, or refer students to pages 7 and 8 in the student edition.

> When you describe a set of data, it is helpful to give the value that occurs most often. Look back at the data for Ms. Jeckle's class. Which name length occurred most often?

A name length of 15 letters occurs most often.

> The value that occurs most often in a set of data is called the *mode.* The mode for Ms. Jeckle's data is 15 letters.
>
> How can you find the mode by looking at a line plot or a bar graph?

On a line graph, the mode has the highest stack of Xs. On a bar graph, the mode has the tallest bar.

> Notice that the shortest name in Ms. Jeckle's class has 9 letters and the longest has 17 letters. The *range* of a data set is the spread of the values from the highest value to the lowest value. The range of Ms. Jeckle's data is from 9 letters to 17 letters.

For the Teacher

Be careful how students word their descriptions of the mode. For example, for Ms. Jeckle's class, the mode of the data set is 15 letters. Students may be inclined to say that "most students have 15 letters in their names." This is not true. There are 24 students in the data set, and only 6 of the students have 15 letters in their names.

The range can be described in more than one way. It can be given by specifying the lowest and the highest values (for example, the range for Ms. Jeckle's class is from 9 letters to 17 letters). The range can also be given by determining the number of values from the lowest to highest value. For Ms. Jeckle's data, the range spreads across 9 name lengths (9, 10, 11, 12, 13, 14, 15, 16, 17). The range is also commonly given by subtracting the lowest value from the highest value. For Ms. Jeckle's data, the range would be 17 − 9, or 8, letters.

In this problem, students are given the number of values in a data set and the mode and range of the values. They must work backward from this information to create a possible distribution. Read aloud the description of the hypothetical class given in Problem 1.3, and ask students questions about the information.

What facts do we know about this class of students?

How can we use this information to come up with a set of name lengths for this class?

Is there more than one possible set of name lengths that fits this description? Why or why not?

Students should see that there are many data sets that fit the description given. Each possible data set must have 15 values between (and including) 8 letters and 16 letters, and the value 12 letters must occur the greatest number of times.

Explore

Have students work in pairs to investigate the problem. Each pair should create a distribution— a line plot—that meets the criteria of the problem and can be posted for the rest of the class to see. You may want to have students make their line plots on large sheets of paper—using stick-on notes instead of Xs—or use transparencies and transparency markers to make graphs that can easily be erased and redrawn.

Summarize

The goal is not for students to determine all possible answers, but to realize that many solutions are possible. The criteria provide some constraints on the shape of the data, but the remaining information can be chosen in many different ways to create a data set.

Have students display their line plots. They will have created a variety of data sets of name lengths. Any graph that satisfies the conditions of the problem is acceptable. See page 21i for some possible graphs.

Look at the different line plots that were made. How are they alike?

All graphs should show 15 Xs, a mode of 12 letters, and a range of 8 letters to 16 letters.

How are the line plots different?

The shapes of the displays may look very different from one another.

Why are different displays possible?

We are not given enough information to create a unique data set. We can fill in the missing information in many different ways.

1.4 • Identifying the Median

Launch

In this section, the median is introduced as a type of average. Students have probably heard the word *average* used most often in the context of "add up all the numbers and divide by the number of numbers." This procedure gives the mean, which is only one type of average. The mean, median, and mode are all types of averages, or *measures of center.* Ideally, when hearing or reading the word *average* with respect to a set of data, students will want to know which "average" is being used.

> In the last problem, you learned to use the mode to describe what is typical or average about a set of data. In this problem, you will find another type of average called the *median.*
>
> Averages such as the median and mode are sometimes referred to as *measures of center.* This name is probably most fitting for the median, because it is the middle point of a set of data.

For the Teacher

The *mode* is the value that occurs with greatest frequency in a set of data. The *median* is the value that separates an ordered set of data in half, with half of the data falling before the median and half falling after the median. Although there is always only one median in a set of data, there may be more than one mode.

Have students refer to the data for Problem 1.4 in their books, or display Transparency 1.4A.

> This table and the line plot show name-length data for a class in Michigan.
>
> We want to find the median, or middle value, for this set of data. Half of the data values will be less than or equal to the median, and half of the values will be greater than or equal to the median. To find the middle value, we must first order the data from smallest to largest.

Write each name length on a large stick-on note, and display the lengths in the order they occur in the table.

Work with the class to rearrange the name lengths in order from smallest to largest.

After you have ordered the data, have pairs work on Problem 1.4.

Explore

Circulate while students work on the problem. If some students finish early, ask them to think about the problem posed in the "Think about this!" box on page 13 of the student edition.

Summarize

To help students understand the median, you may want to create your own large strips of ordered data values, one strip for the original 21 values and the other for the 22 values. In each case, draw a dark line to indicate the median. You could use gridded chart paper (available in office-supply stores). Hold up the strip with 21 values.

> For the strip of 21 values, the fold landed on a 12. Half the values were before the fold and half were after the fold. We say that the *median* name length for this set of names is 12 letters.

Now hold up the strip of 22 values.

> For the strip of 22 values, the fold did not land on a number. It fell between 12 and 13. In a case like this, where there is an even number of values, we define the *median* as the number halfway between the two middle values. In this case, the median is $12\frac{1}{2}$ letters.

Now, direct students to the table and graph of the original data.

> Look back at the original data set and line plot. The median name length is 12 letters. Where is this located on the line plot?

It is one of the values above the 12 on the line plot.

> Earlier we found a different kind of average called the *mode*. How does the median compare with the mode in this data set?

The data set has two modes, 11 letters and 12 letters. The median—12 letters—is the same as one of the modes.

> After we add the data for Suzanne Mannerstrale, how does the median compare with the mode?

The median ($12\frac{1}{2}$ letters) is no longer the same as either of the modes. You may want to point out that, while the mode is always a value in the data set, the median may not be. In this case, the median, $12\frac{1}{2}$ letters, is not even a possible value for a name length.

> What would happen to the median and the mode if another student joined the class with a name length of 11 letters?

The median would be 12 letters, and the mode would be 11 letters.

> Can the median be the same as the mode?

The two measures of center can have the same value.

Folding a strip of paper to find the median would be an inefficient strategy if we had a lot of data values. Can you think of another way to find the median?

Students may have several valid suggestions. One possibility is to make an ordered list of the name lengths and count in from the ends, pairing one length from the "small" end with one length from the "large" end until you work your way to the middle. Another possibility is to make an ordered list of the name lengths. If there is an even number of data values, divide the number of values in half. The result will be a whole number. Count that many values from one end of the data. The median will be halfway between the number you land on and the next number. If there is an odd number of data values, divide the number of values in half. The result will be a fraction. Round the fraction to the next whole number, and count over that many values from one end of the data. The number you land on will be the median.

Have students work on Problem 1.4 Follow-Up at home or during class time. The questions will give them additional practice with finding the median and using it to describe what is typical about a set of data. Be sure to discuss the answers in class.

For the Teacher

In a data set with an even number of values, where the two middle values differ by more than one, the median is still the value halfway between these values. For example, the median for the data 3, 4, 4, 7, 8, 9 is $5\frac{1}{2}$, the number halfway between 4 and 7. If the two middle values are the same number, the median is that number. For example, for the data 3, 4, 5, 5, 7, and 8, the median falls between the two 5s, so the median is 5.

1.5 • Experimenting with the Median

Launch

In this problem, students explore the stability of the median: how responsive is the median to changes in the data? Does it change if we add a very large or a very small value to the data? How does it react if we make other changes in the data? The idea of the median and its stability is important in making judgments about statistical data.

Each group of students should prepare ten index cards as described in the student edition. Once they have ordered the names from shortest to longest, have them determine the median ($11\frac{1}{2}$ letters). Then, work with the class to find an example for part A.

> Let's look at part A of Problem 1.5. Starting with the ten cards, we need to see if we can take away two cards without changing the median. Which cards might we choose?

You could, for example, remove the lowest card and the highest card.

When students understand the nature of the questions in the problem, let them work in small groups on Problem 1.5. Ask them to try to find at least three different possibilities for each part. Remind them to keep a record of what they find out.

Explore

While students are working on the problem, take time to work with those that are having difficulty. Work through a conjecture they suggest and see how it changes the median.

Summarize

For each part of the problem, have one or two groups present examples that meet the criteria. You want students to realize that the median is a fairly stable value. It doesn't change significantly when a few additional values—no matter how large or small—are added to the data set. They will later see that this important characteristic makes the median a useful number to statisticians.

For the Teacher

Use part D of Problem 1.5 and Problem 1.5 Follow-Up to point out that adding a very large number has little effect on the median. In part D, the median would increase to 12 letters whether we added name lengths of 16 letters and 17 letters or 900 letters and 1000 letters. In the follow-up, adding a name length of 1019 letters to the data has the same effect on the median as adding a name length of 16 letters.

Additional Answers

Answers to Problem 1.3

B.   Possible line plots:

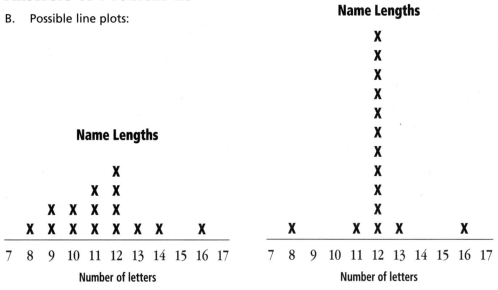

ACE Questions

Applications

1.

Name	Letters
Ben Carter	9
Ava Baker	8
Sarah Edwards	12
Juan Norinda	11
Ron Weaver	9
Bryan Wong	9
Toby Vanhook	11
Katrina Roberson	15
Rosita Ramirez	13
Kimberly Pace	12
Paula Wheeler	12
Darnell Cox	10
Jessica Otto	11
Erin Foryeh	10
Corey Buysse	11
Tijuana Degraffenreid	20

Name Lengths

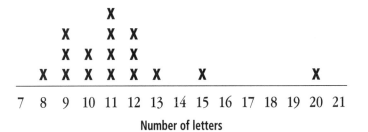

Name Lengths

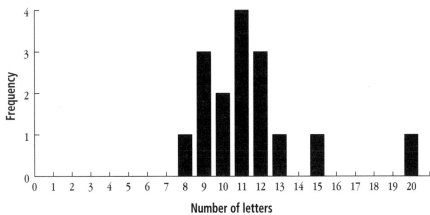

8. Possible answer:

Name Lengths

```
            X   X
    X   X   X   X           X
  ─────────────────────────────────
  11  12  13  14  15  16  17  18  19
```
Number of letters

9. Possible answer:

Name Lengths

```
                    X
                    X
              X   X   X   X
      X       X   X   X   X   X           X
    ─────────────────────────────────────────
    7   8   9   10  11  12  13  14  15  16  17  18
```
Number of letters

10. Possible answer:

Name Lengths

```
        X
        X       X
    X   X   X   X           X       X
    X   X   X   X       X   X       X
  ───────────────────────────────────────
  10  11  12  13  14  15  16  17  18  19  20  21
```
Number of letters

Mathematical Reflections

4. The mode and the median for a set of data may or may not be the same. For the data, 1, 2, 3, 3, 4, 5, 6, both the median and the mode are 3. For the data set, 1, 1, 1, 3, 5, 6, 6, the mode is 1 and the median is 3.

5. The range indicates how spread out the data are. Combined with a measure of center such as the median or the mode, the range helps to give a picture of the data. For example, if you know a data set has a median of 20, you know where the middle of the data set is. If, in addition, you know the range is from 18 to 22 (or from 1 to 60), you have a much better idea of what the data may look like.

6. The shape of the data includes such things as clusters, peaks, and gaps in the data. Bar graphs and line plots help you see the shape of the data.

7. The mode, median, and range can be used to describe what is typical about a data set. You can also give an interval in which most of the data values fall. Giving information about the shape of the data (peaks, gaps, clusters) also helps describe what is typical.

Types of Data

Mathematical and Problem-Solving Goals

- **To note the kind of data being collected; that is, categorical or numerical**

- **To use bar graphs to display categorical and numerical data**

- **To understand how measures of center (median, mode) and spread (range) relate to numerical and categorical data**

Questions in real life concern two kinds of data: categorical and numerical. Knowing the type of data helps us to determine the most appropriate measures of center and displays to use for the data.

When we collect data, we are collecting a "measurement" about some "thing." We are interested in organizing the data by tallying, or finding the frequency of occurrence for each data value. This investigation helps students to classify data values as categorical or numerical. Several examples are given in the student edition to help illustrate the distinction between these types of data.

In Problem 2.1, Category and Number Questions, students write questions with numerical answers and questions with categorical answers. In Problem 2.2, Counting Pets, students investigate categorical and numerical data about pets and determine whether given questions can be answered using the data given, or whether more information is required.

Note: Problem 3.1 asks students to collect data on their travel times to school. We recommend you read the teacher's edition notes for Problem 3.1 and have students start to collect this data while you are doing Investigation 2.

Student Pages	22–29
Teaching the Investigation	29a–29e

Materials

For students
- Chart paper (optional)

For the teacher
- Transparencies 2.1, 2.2A, and 2.2B (optional)

2.1

Category and Number Questions

Launch

- Have students focus on the questions posed in the student edition and consider the types of answers each would generate.

- Help students summarize what they know about categorical and numerical data.

Explore

- Circulate while students work, observing any interesting questions you want shared with the class.

Summarize

- As a class, talk about the questions students generated.

- Start a class list of numerical and categorical questions. (*optional*)

- Have students think about whether the data generated by their questions is what they intended.

Assignment Choices

ACE questions 1–8 and unassigned choices from earlier problems

---

Types of Data

When we are interested in finding out more about something, we start asking questions about it. Some questions have answers that are words or categories, for example, What is your favorite sport? Other questions have answers that are numbers, for example, How many inches tall are you?

Read each of the questions below. Which questions have words or categories as answers? Which questions have numbers as answers?

- In what month were you born?
- What kinds of pets do you have?
- How many pets do you have?
- Who is your favorite author?
- How much time do you spend watching television in a day?
- What's your highest score in the game Yahtzee?
- What color are your eyes?
- How many movies have you watched in the last week?
- How do you get to school?

2.1 Category and Number Questions

The data you collect in response to a question you ask may be numbers or words.

Data that are words or categories are called **categorical data.** Categorical data are usually not numbers. If you asked people in which month they were born or what kinds of pets they have, their answers would be categorical data.

Data that are numbers are called **numerical data.** If you asked people how tall they are or how many pets they have, their responses would be numerical data.

Problem 2.1

Think of some things you would like to know more about. Then, develop some questions you could ask to gather information about those things.

A. Write two questions that have categorical data as answers.

B. Write two questions that have numerical data as answers.

▓ Problem 2.1 Follow-Up

Is it possible to find the mode of a set of categorical data? Explain your answer.

2.2 Counting Pets

The pets people have often depend on where they live. People who live in cities often have small pets, while people who live on farms often have large pets. People who live in apartments are sometimes not permitted to have pets at all.

It is fun to find out what kinds of pets people have. One middle-school class gathered data about their pets by tallying students' responses to these questions:

What is your favorite kind of pet?

How many pets do you have?

The students' questions produced two kinds of data. When students told what their favorite pets were, their responses were categorical data. When students told how many pets they had, their responses were numerical data.

At a Glance

Launch

■ As a class, inspect the data that the students in Problem 2.2 collected.

■ Ask questions to help students read the tables and the graphs.

Explore

■ Circulate while students work on the questions about the data.

Summarize

■ As a class, talk about students' responses to the questions.

■ Have students write additional questions that can and cannot be answered by the given data, and discuss their questions. (*optional*)

Answers to Problem 2.1

Answers will vary.

Answers to Problem 2.1 Follow-Up

yes; You could figure out which category occurs most frequently.

Assignment Choices

ACE questions 9–17, and unassigned choices from earlier problems

Assessment

It is appropriate to use Check-Up 1 after this problem.

The students made tables to show the tallies or frequencies, and then made bar graphs to display the data.

Favorite Kinds of Pets

Pet	Frequency
cat	4
dog	7
fish	2
bird	2
horse	3
goat	1
cow	2
rabbit	3
duck	1
pig	1

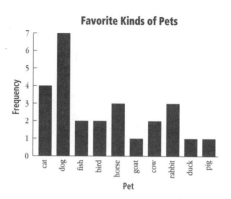

Favorite Kinds of Pets

Numbers of Pets

Number of pets	Frequency
0	2
1	2
2	5
3	4
4	1
5	2
6	3
7	0
8	1
9	1
10	0
11	0
12	1
13	0
14	1
15	0
16	0
17	1
18	0
19	1
20	0
21	1

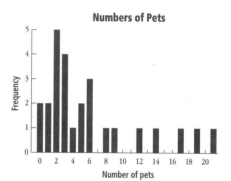

Numbers of Pets

Answers to Problem 2.2

A. The graph of favorite pets shows categorical data. The graph of number of pets shows numerical data.

B. 132 pets (multiply each number of pets by its frequency, then add the results)

C. 21 pets (this is the highest number in the "Number of Pets" column of the "Numbers of Pets" table)

D. 26 students (add the numbers in the "Frequency" column of either table or add the heights of the bars in either graph.)

E. 4 students

F. This question cannot be answered from the data given.

(Continued on next page.)

Problem 2.2

Decide whether each question below can be answered by using data from the graphs the students created. If a question can be answered, give the answer and explain how you got it. If a question cannot be answered, explain why not and tell what additional information you would need to answer the question.

A. Which graph shows categorical data, and which graph shows numerical data?

B. What is the total number of pets the students have?

C. What is the greatest number of pets that any student in the class has?

D. How many students are in the class?

E. How many students chose cats as their favorite kind of pet?

F. How many cats do students have as pets?

G. What is the mode for the favorite kind of pet?

H. What is the median number of pets students have?

I. What is the range of the numbers of pets students have?

J. Tomas is a student in this class. How many pets does he have?

K. Do the girls have more pets than the boys?

▓ Problem 2.2 Follow-Up

Do you think the students surveyed live in a city, the suburbs, or the country? Explain your answer.

G. dogs

H. $3\frac{1}{2}$ pets

I. The range is 0 pets to 21 pets.

J. This question can't be answered from the data. Data by individual students were not collected.

K. This question can't be answered from the data. Data by gender were not collected.

Answer to Problem 2.2 Follow-Up

Let your students make conjectures about where the data were collected, making sure they justify their responses. After they have discussed their ideas, you may want to reveal that the students who actually provided these data live in rural North Carolina.

Answers

Applications

1. numerical

2. categorical

3. numerical

4. categorical

5. Although this question does not have a numerical answer, numerical data must be collected to answer it.

6. numerical

7. categorical

8. numerical

Connections

9a. Half of all rats live less than $2\frac{1}{2}$ years, and half live longer than $2\frac{1}{2}$ years.

9b. Knowing the range in ages for rats would be helpful.

As you work on these ACE questions, use your calculator whenever you need it.

Applications

In 1–8, tell whether the answers to the question are numerical or categorical data.

1. What is your height in centimeters?

2. What is your favorite musical group?

3. On a scale of 1 to 7, with 7 being outstanding and 1 being poor, how would you rate the food served in the school cafeteria?

4. What would you like to do when you graduate from high school?

5. Are students in Mr. P's class older in months than students in Ms. J's class?

6. How many of your own feet tall are you?

7. What kind(s) of transportation do you use to get to school?

8. How much time do you spend doing homework?

Connections

9. Alicia has a rat that is three years old. She wonders if her rat is old compared to other rats. At the pet store, she finds out that the median age for a rat is $2\frac{1}{2}$ years.

a. What does the median tell Alicia about the life span for a rat?

b. What additional information would help Alicia predict the life span of her rat?

In 10–13, use the graph below, which shows the numbers of sodas consumed by 100 middle-school students in one day.

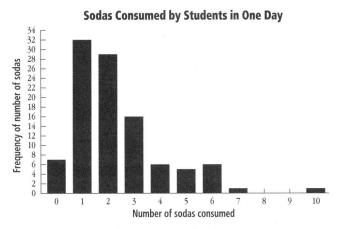

Sodas Consumed by Students in One Day

10. Are these data numerical or categorical? Explain your answer.

11. A student used this graph to estimate that the median number of sodas drunk by these students in a day was 5. Is this student correct? Explain your answer.

12. Another student estimated that the median number of sodas drunk in a day was 1. Is this student correct? Explain your answer.

13. What is the total number of sodas drunk by the 100 students in one day? Describe how you determined your answer.

14. Suppose these students were asked, What kinds of soda do you like to drink?

    a. Give three possible responses to this question.

    b. Describe how you would make a bar graph to show the data that would be collected to answer this question. What would the label for the horizontal axis be? What would the label for the vertical axis be? What would the title of the graph be? What would each bar on the graph show?

---

13. The total number of sodas drunk is determined by evaluating each bar of the graph:

    7 people × 0 sodas = 0 sodas   5 people × 5 sodas = 25 sodas

    32 people × 1 soda = 32 sodas   3 people × 6 sodas = 18 sodas

    29 people × 2 sodas = 58 sodas   1 person × 7 sodas = 7 sodas

    16 people × 3 sodas = 48 sodas   1 person × 10 sodas = 10 sodas

    6 people × 4 sodas = 24 sodas

So, 100 people drank a total of 222 sodas in one day.

---

10. numerical; The answer to the question "How many sodas do you drink in one day?" is a number.

11. no; The median is the number that separates the ordered data in half. The number of people that drink 5 sodas in one day is near the upper end of the data, so 5 cannot be the median.

12. no; There are 100 students, so the median is between the fiftieth and fifty-first ordered data value. A total of 39 students drank 0 or 1 soda in one day. This means the median is greater than 1 soda, because the fiftieth value will be in the next column—2 sodas in one day.

13. See below left.

14a. Possible answer: root beer, orange soda, and cherry soda

14b. The graph for the categorical data could have a horizontal axis showing the names of the sodas and labeled "Kind of soda." The vertical axis could display the number of students that said they liked each type of soda and be labeled "Frequency of kind of soda." The title of the graph could be "Kinds of Soda Consumed by Students." Each bar of the graph would show how many people had chosen that particular soda as a soda they liked to drink. The height of the bar would be the frequency of that choice.

Extensions

15. See page 29d.

16. Possible answer: Fish occur the most frequently, followed by cats and dogs. Most of the remaining pets are also "indoor" pets.

17. Answers will vary. Some students may immediately respond that 841 people were surveyed, indicating that each person surveyed had one pet. Other students may note that this response does not take into account that it is likely that some people surveyed had no pets or had more than one pet.

Extensions

In 15–17, use the data below. These data were collected from a large number of middle-school students and show the kinds of pets the students have. Of a total of 841 pets, the table shows that 61 are birds and 184 are cats. From this data we cannot tell how many students were surveyed. We only know that when the survey was completed, a total of 841 pets had been counted.

Kinds of Pets Students Have

Kind of pet	Frequency of kind of pet
bird	61
cat	184
dog	180
fish	303
gerbil	17
guinea pig	12
hamster	32
horse	28
rabbit	2
snake	9
turtle	13
Total	**841**

15. Make a bar graph to display this data. Think about how you will design and label the horizontal and vertical axes.

16. Use the information displayed in your graph to write a paragraph about the pets these students have.

17. What might be a good estimate of how many students were surveyed? Explain your reasoning.

Tips for the Linguistically Diverse Classroom

Original Rebus The Original Rebus technique is described in detail in *Getting to Know CMP*. Students make a copy of the text before it is discussed. During discussion, they generate their own rebuses for words they do not understand as the words are made comprehensible through pictures, objects, or demonstrations. Example: key words for which students may make rebuses are the names of animals that are unfamiliar.

Mathematical Reflections

In this investigation, you explored different kinds of data. These questions will help you summarize what you have learned:

1 How would you explain what categorical and numerical data are to a classmate who missed this investigation?

2 You have learned to use the mode and median to describe what is typical about a set of data. Can the mode or median be used to describe categorical data? Can the mode or median be used to describe numerical data?

3 The range is used to help describe how spread out a set of data are. Can you use the range to describe categorical data? Can you use the range to describe numerical data?

Think about your answers to these questions, discuss your ideas with other students and your teacher, and then write a summary of your findings in your journal.

To carry out a research project about characteristics of the typical middle-school student, you will need to pose questions. What questions might you ask that would have categorical data as answers? What questions might you ask that have numerical data as answers?

Possible Answers

1. *Numerical data* are numbers. For example, if you asked your classmates, "How many months old are you?" the resulting data would be numerical data. *Categorical data* are words or groups. For example, if you asked the question, "What is your favorite book?" the resulting data would be categorical data.

2. The mode—the data value that occurs most frequently—can be used to describe both categorical and numerical data. In the categorical data for Problem 2.2, the mode is "dog," which is the favorite pet that occurs most frequently. In the numerical data for Problem 2.2, the mode is 2, the number of pets that occurs most frequently.

The median—the center of an ordered set of data—can only be used to describe numerical data. In the data for Problem 2.2, it does not make sense to talk about the median type of pet, because it is not possible to list the kinds of pets in an ordered sequence.

3. There can be no range for categorical data, because the values of a specific measure (such as kind of pets) cannot be ordered. The range means something only with data that can be ordered from smallest to largest. This is only true of numerical data.

Tips for the Linguistically Diverse Classroom

Original Rebus The Original Rebus technique is described in detail in *Getting to Know CMP*. Students make a copy of the text before it is discussed. During discussion, they generate their own rebuses for words they do not understand as the words are made comprehensible through pictures, objects, or demonstrations. Example: Question 1—key words for which students may make rebuses are *categorical* (three different kinds of flowers), *numerical* (three different numbers), *classmate who missed this investigation* (a stick figure in bed with a thermometer in its mouth).

TEACHING THE INVESTIGATION

2.1 • Category and Number Questions

Launch

Refer students to the questions posed on the top of page 22 in the student edition.

> Look over these questions, and think about how you would respond to each. To which questions would you respond with a word? To which questions would you respond with a number?

Discuss the kinds of responses students would make. This may provoke some lively discussion about how the questions should be interpreted and how the data might be collected. Let some of this discussion occur. The goal is to have students think about the *kinds of responses,* not to actually gather data. Here are a few more examples of numerical and categorical data:

Numerical data

- We can collect data about family size and organize them by tallying how many families have zero children, one child, two children, and so on.

- We can collect data about pulse rates and organize them into intervals by tallying how many people have pulse rates in the interval of 60–69 beats, 70–79 beats, and so on.

- We can collect data about height and organize them into intervals by tallying how many people are between 40–44 inches tall, 45–49 inches tall, and so on.

- We can collect data about time spent sleeping in one day and organize them by tallying how many people slept 7 hours, $7\frac{1}{2}$ hours, 8 hours, and so on.

- We can collect data about responses to a question such as, "On a scale of 1 to 5 with 1 as 'low interest,' rate your interest in participating in the school's field day" and organize them by tallying how many people indicated each of the ratings 1, 2, 3, 4, or 5.

Categorical data

- We can collect data about birth years and organize them by tallying how many people were born in 1980, 1981, 1982, and so on.

- We can collect data about favorite type of book to read and organize them by tallying how many people like mysteries, adventure stories, science fiction, and so on.

- We can collect data about hobbies and organize them by tallying how many people collect stamps, build models, make jewelry, and so on.

Help students to summarize what they know about the distinction between categorical and numerical data, and then let them work on Problem 2.1.

Explore

Circulate among students, helping them if necessary and looking for interesting questions to be shared in the summary.

Summarize

It might be helpful to begin a display like the one shown below, grouping students' questions into categories.

Questions with Categorical Data Answers
What kinds of pets do you have?
In what month were you born?
Who is your favorite author?

Questions with Numerical Data Answers
How many pets do you have?
How much time do you spend watching television in a day?
What's your highest score in the game Yahtzee?
How many movies have you watched in the last week?

Have students write their questions on long strips of paper and post them in one of the columns. Students may decide to use these questions for their final Is Anyone Typical? project.

Encourage students to think about the kinds of data their questions will elicit and about ways in which their questions might be misinterpreted, resulting in data that is not what they intended. Students can imagine some responses and think about whether the responses actually answer the questions in which they are interested.

2.2 • Counting Pets

Launch

The goals of Problem 2.2 are to help students think about categorical and numerical data and to give students practice in reading information from graphs. We recommend that you launch the problem in two stages. In the first stage, help your students understand why they cannot find the median or range of categorical data. During this stage, students should work with their books closed.

One teacher conducted the first stage using the pet data from page 24 of the student edition. She gave each group a copy of the data listed in a *different order* from the copies she gave to the other groups. This made it likely that students would list the pets in different orders along the horizontal axes of their "favorite pet" graphs. This is how she began the discussion:

> A class gathered data about their pets. Here are the two questions that each student answered. *(Write these on the board or overhead.)*
>
> ■ What is your favorite kind of pet?
> ■ How many pets do you have?
>
> The sheets I just handed you show the data the students collected for each question. For each set of data, I would like you to make a graph and then find the range, median, and mode. Make your graphs large enough so the whole class will be able to see them.

When the groups displayed their results, they found that, although they all found the same mode for the favorite pets, each group found a different range and median. This was because they were using the favorite pet data as if it were ordered. This helped students to understand the fundamental difference between categorical data and numerical data. In the "favorite pet" data, which is categorical, there is no logical way to order the data, so no median or range can be found. In the "how many pets" data, the values can be ordered numerically, and a median and range can be calculated. When the teacher saw that her students understood this difference, she had them open their books to page 23 and proceed with Problem 2.2.

For the Teacher

At times, categorical data seem to be organized like numerical data. For example, a bar graph of birth months may employ numbers to represent months (1 for January, 2 for February, 3 for March). However, we cannot perform numerical operations using months of the year, because months are categorical data. Numerical measures such as median, mean, and range cannot be used to interpret categorical data.

Refer students to page 24, which shows the tables and bar graphs (the graphs also appear on Transparency 2.2A). Work with students to make sure they can read the two tables and understand the data as it is presented. Discuss how to read the graphs, highlighting the information shown on the horizontal axis and the vertical axis for each graph. You may want to ask a few questions to sharpen table and graph comprehension skills. (It is all right for students to move back and forth between the table and the graph. Just make sure they can locate information using both representations.) You can ask questions, such as those below, to help students to read the data, to read between the data, and to read beyond the data.

Reading the data

Ask how many students chose a dog as their favorite pets (7) and how many students have 6 pets (3).

Reading between the data

Ask how many total students chose dogs or cats as their favorite pets (11), and how many people have more than 6 pets (7).

Reading beyond the data

Ask students what they know about the kinds of pets these students chose as their favorites. (Answers will vary. The most popular pet—the mode—is the dog, and the second most popular is the cat. The presence of cow, goat, duck, and pig may suggest that some of the students live on farms.)

Ask students what they know about how many pets each of the students has. (Answers will vary. The range is from 0 to 21 pets. Only 2 people do not have pets. The mode is 2. The data cluster in the interval of 2–6 pets, then spread out with some unusual values occurring in the interval from 14–21 pets.)

Do you think our class data would be similar to or different from these students' data?

Explore

Let students work in small groups to discuss their responses to each of the Problem 2.2 questions.

Summarize

Have a class discussion in which teams of students explain their responses to the questions. It is important for students to understand what they can and cannot know from a set of data.

To complete this activity, you may want students to work in pairs for about five minutes, seeing if they can write some questions about these data that can and cannot be answered. For questions that cannot be answered, discuss what information would be needed to answer them.

Additional Answers

ACE Questions

Extensions

15. Possible answer:

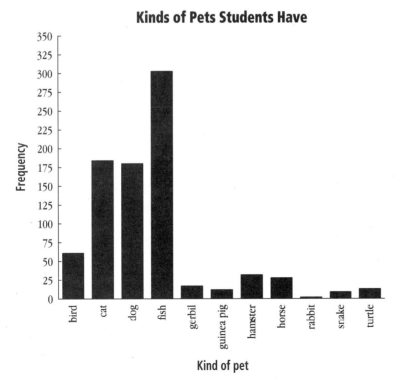

Kinds of Pets Students Have

The challenge for students will be developing the scale for the vertical axis. Because of the range of the data (9 to 303 pets), the scale probably needs to be numbered by at least fives or tens.

Using Graphs to Group Data

In this investigation, students explore two different contexts in which the data collected are more variable than those in earlier data sets. Representations such as line plots and bar graphs are not suitable for displaying these data; the patterns within the data sets can only be seen when the data are grouped. The stem-and-leaf plot provides a representational tool that groups data in intervals.

In Problem 3.1, Traveling to School, students use a stem-and-leaf plot to investigate the time it takes each student in a middle-school class to travel to school. In Problem 3.2, Jumping Rope, students use a back-to-back stem plot to compare jump-rope endurance for two classes.

Mathematical and Problem-Solving Goals

- *To use stem-and-leaf plots to group numerical data in intervals*

- *To use the ordered data in a stem plot to locate measures of center (median and mode) and measure of spread (range)*

- *To describe the shape of the data, including the location of clusters and gaps, and to determine what is typical about the data*

- *To compare two data sets by using back-to-back stem-and-leaf plots*

- *To compare two data sets by using statistics, such as median and range*

Materials

For students
- Graph paper

For the teacher
- Transparencies 3.1A, 3.1B, and 3.2 (optional)
- Local street map (optional)

Student Pages   30–41

Teaching the Investigation   41a–41i

3.1

Traveling to School

At a Glance

Launch

■ With students, analyze the travel-time data in the student edition, and consider whether a line plot would be a useful representation for the data.

Explore

■ As a class, construct a stem plot for the data.

Summarize

■ Help the class read the stem plot and identify intervals.

■ As a class, collect and organize data about how long it takes students to get to school and (optionally) how far they travel and their mode of travel.

■ Help students to "read beyond" the data they have collected.

Using Graphs to Group Data

The data you have seen so far have had small ranges. For example, the name lengths in Problem 1.2 ranged from 9 to 17 letters, and the numbers of pets in Problem 2.2 ranged from 0 to 21 pets. You could see the shape of these data by examining a bar graph or a line plot. For data with a large range, a bar graph or a line plot, with a bar or stack of Xs for every value, often does not give a good idea of the shape of the data. In this investigation, you will learn about a type of graph that groups data values into intervals, making it easier to see the shape of the data.

3.1 Traveling to School

While investigating the times they got up in the morning, a middle-school class in Wisconsin was surprised to find that two students got up almost an hour earlier than their classmates. These students said they got up early because it took them so long to get to school. The class then wondered how much time it took each student to travel to school in the morning. The data they collected are on the next page.

Notice that the data about distances are recorded in decimal form: 4.50 miles means the same thing as $4\frac{1}{2}$ miles. What fractions would you write for 0.75 miles and 2.25 miles?

Think about this!

Look at the table of data and the labels for the columns, and consider these questions.
• What three questions did the students ask?
• How might the students have collected the data to answer these questions?
• Would a line plot be a good way to show the travel-time data? Why or why not?

---

30 Data About Us

Assignment Choices

ACE questions 1–7 and unassigned choices from earlier problems

Tips for the Linguistically Diverse Classroom

Rebus Scenario The Rebus Scenario technique is described in detail in *Getting to Know CMP*. This technique involves sketching rebuses on the chalkboard that correspond to key words in the story or information you present orally. Example: some key words and phrases for which you may need to draw rebuses while discussing the material on this page: *times* (several morning times indicated on a digital clock), *two students got up almost an hour earlier* (stick figures getting out of bed with a clock showing the time) *than their classmates* (stick figures sleeping and a clock showing a time one hour later than the time on the first clock), *took them so long to get to school* (original stick figures arriving at school and a clock indicating a time much later than the time they woke up).

---

30 Investigation 3

Student's initials	Time (minutes)	Distance (miles)	Mode of travel
DB	60	4.50	bus
DD	15	2.00	bus
CC	30	2.00	bus
SE	15	0.75	car
AE	15	1.00	bus
FH	35	2.50	bus
CL	15	1.00	bus
LM	22	2.00	bus
QN	25	1.50	bus
MP	20	1.50	bus
AP	25	1.25	bus
AP	19	2.25	bus
HCP	15	1.50	bus
KR	8	0.25	walking
NS	8	1.25	car
LS	5	0.50	bus
AT	20	2.75	bus
JW	15	1.50	bus
DW	17	2.50	bus
SW	15	2.00	car
NW	10	0.50	walking
JW	20	0.50	walking
CW	15	2.25	bus
BA	30	3.00	bus
JB	20	2.50	bus
AB	50	4.00	bus
BB	30	4.75	bus
MB	20	2.00	bus
RC	10	1.25	bus
CD	5	0.25	walking
ME	5	0.50	bus
CF	20	1.75	bus
KG	15	1.75	bus
TH	11	1.50	bus
EL	6	1.00	car
KLD	35	0.75	bus
MN	17	4.50	bus
JO	10	3.00	car
RP	21	1.50	bus
ER	10	1.00	bus

Investigation 3: Using Graphs to Group Data 31

The students decided to make a *stem-and-leaf plot* of the travel times.

Making a Stem-and-Leaf Plot
A **stem-and-leaf plot** looks something like a stem with leaves. It is sometimes simply called a *stem plot*.

To make a stem plot, begin by looking at the data values. Ignore the ones digits and look at the remaining digits of the numbers. These digits will make up the "stem." For these data, the stem will be made up of the tens digits. Since the travel times range from 5 minutes to 60 minutes, the stem will be made up of the digits 0, 1, 2, 3, 4, 5, and 6. Make a vertical list of the tens digits in order from smallest to largest, and draw a line to the right of the digits to separate the stem from the "leaves."

```
0 |
1 |
2 |
3 |
4 |
5 |
6 |
```

The "leaves" are the ones digits. For each data value, you add a leaf next to the appropriate tens digit on the stem. For example, the first data value is 60 minutes. You show this by writing a 0 next to the stem value of 6. The next value is 15 minutes. Indicate this by writing a 5 next to the stem value of 1.

```
0 |
1 | 5
2 |
3 |
4 |
5 |
6 | 0
```

The next few travel times are 30 minutes, 15 minutes, 15 minutes, 35 minutes, 15 minutes, 22 minutes, 25 minutes, and 20 minutes. Can you figure out how these data were added as leaves to the stem plot?

```
0 |
1 | 5 5 5 5
2 | 2 5 0
3 | 0 5
4 |
5 |
6 | 0
```

Copy and complete the stem plot, and compare it to the one below.

```
0 | 8 8 5 5 5 6
1 | 5 5 5 5 9 5 5 7 5 0 5 0 5 1 7 0 0
2 | 2 5 0 5 0 0 0 0 0 1
3 | 0 5 0 0 5
4 |
5 | 0
6 | 0
```

After you have added leaves for all the data values, redraw the plot, listing the ones digits in order from smallest to largest. Then, add a key showing how to read the plot and give the plot a title. Compare your new stem plot to the one below.

Travel Times to School (minutes)

```
0 | 5 5 5 6 8 8
1 | 0 0 0 0 1 5 5 5 5 5 5 5 5 5 7 7 9
2 | 0 0 0 0 0 0 1 2 5 5
3 | 0 0 0 5 5
4 |
5 | 0                        Key
6 | 0            2 | 5   means 25 minutes.
```

3.2

Jumping Rope

Launch

- Decide whether to have students collect their own jump-rope data or use the given data sets.

- Help students to read the back-to-back stem plot.

Explore

- Arrange for students to collect their own data. (*optional*)

- Have students work in groups on the problem and follow-up.

Summarize

- In a class discussion, talk about ways to compare the two sets of data.

Problem 3.1

Read "Making a Stem-and-Leaf Plot" to explore how to make a stem-and-leaf plot of the travel-time data. After you have completed your stem plot of the data, answer these questions.

A. Which students probably get to sleep the latest in the morning? Why do you think this?

B. Which students probably get up earliest? Why do you think this?

C. What is the typical time it takes for these students to travel to school?

■ Problem 3.1 Follow-Up

Consider this question: What is the typical time it takes for a student in your class to travel to school?

1. Decide what data you need to collect to answer this question. Then, with your classmates, gather the appropriate data.

2. Find a way to organize and display your data.

3. After looking at your data, what would you say is the typical time it takes for a student in your class to travel to school?

3.2 Jumping Rope

While doing a jump-rope activity in gym class, a student in Ms. Rich's class wondered what was the typical number of jumps a middle-school student could make without stopping. The class decided to explore this question by collecting and analyzing data. After a practice turn, each student jumped as many times as possible, while a partner counted the jumps and recorded the total. When Mr. Kocik's class found out about the activity, they wanted to join in too.

The classes made a *back-to-back stem plot* (shown on the next page) to display their data. Look at this plot carefully, and try to figure out how to read it.

When the two classes compared their results, they disagreed about which class did better. Mr. Kocik's class pointed out that the range of their data was much greater. Ms. Rich's class said this was only because they had one person who jumped many more times than anybody else. They claimed that most of the students in their class jumped more times than most of the students in Mr. Kocik's class. Mr. Kocik's class disagreed, saying that, even if they did not count the person with 300 jumps, they still did better.

Assignment Choices

ACE questions 8 and 9 and unassigned choices from earlier problems

Answers to Problem 3.1

A. Students who have the shortest travel times—times in the 0–9 minute interval or 10–19 minute interval—probably sleep the latest.

B. Students who have longer travel times—times in the 20–29 or 30–39 minute intervals, and the 50 minute and 60 minute outliers—probably get up early.

C. Answers will vary. Students may want to talk about the intervals in which the data cluster; most of the students are in the intervals of 10–19 minutes and 20–29 minutes. Students may want to find the median, which is 16 minutes.

Answers to Problem 3.1 Follow-Up

Answers will vary.

Numbers of Jumps

Ms. R's class		Mr. K's class
8 7 7 7 5 1 1 1	0	1 1 2 3 4 5 8 8
6 1 1	1	0 7
9 7 6 3 0 0	2	3 7 8
5 3	3	0 3 5
5 0	4	2 7 8
	5	0 2 3
2	6	0 8
	7	
9 8 0	8	
6 3 1	9	
	10	2 4
3	11	
	12	
	13	
	14	
	15	1
	16	0 0
	17	
	18	
	19	
	20	
	21	
	22	
	23	
	24	
	25	
	26	
	27	
	28	
	29	
	30	0

Key

7 | 3 | 0   means 37 jumps for
Ms. R's class and
30 jumps for Mr. K's class.

Answer to Problem 3.2

There are a variety of ways students can respond to this question. They may compute and compare the medians and the ranges, discuss the presence of outliers, or describe the shape of the data. The median number of jumps for Ms. Rich's class is $26\frac{1}{2}$, with a range of 1 to 113 jumps. The median number of jumps for Mr. Kocik's class is 34, with a range of 1 to 300 jumps. If the four outliers in Mr. Kocik's class are ignored, the median number of jumps is 29, with a range of 1 to 104 jumps. In the end, students need to give a well-developed response that makes their reasoning clear.

Problem 3.2

Which class did better overall in the jump-rope activity? Use what you know about statistics to help you justify your answer.

▧ Problem 3.2 Follow-Up

In Mr. Kocik's class, there are some very large numbers of jumps. For example, one student jumped 151 times, and another student jumped 300 times. We call these data *outliers*. **Outliers** are values that stand out in a set of data.

1. Find two other outliers in the data for Mr. Kocik's class.

Statisticians question outliers and try to figure out why they might have occurred. An outlier may be a value that was recorded incorrectly, or it may be a signal that something special is happening that you may want to understand.

2. All the values recorded for Mr. Kocik's class are correct. What do you think might account for the few students who were able to jump many more times than their classmates?

In 3–5, use the data you collected in Problem 3.1 Follow-Up about the time it takes for you and your classmates to travel to school.

3. Make a back-to-back stem plot showing your class data and the data from the Wisconsin class that was used in Problem 3.1.

4. How do your data and the Wisconsin data compare?

5. Are there any outliers in either of the two data sets? Explain.

Answers to Problem 3.2 Follow-Up

1. The two other outliers in Mr. Kocik's class are both 160 jumps.

2. Answers will vary.

3. Students may use the stem plot they made earlier to show travel times for Wisconsin students. They can add their data as leaves on the left side, creating a back-to-back stem plot.

4. Answers will vary.

5. Answers will vary. In the Wisconsin class data, the outliers are 50 minutes and 60 minutes.

Applications • Connections • Extensions

As you work on these ACE questions, use your calculator whenever you need it.

Applications

In 1–4, use this stem-and-leaf plot, which shows the number of minutes it took a class of students to travel to school.

Travel Times to School (minutes)

```
0 | 3 3 5 7 8 9
1 | 0 2 3 5 6 6 8 9
2 | 0 1 3 3 3 5 5 8 8
3 | 0 5
4 | 5
```

Key

2 | 5 means 25 minutes.

1. How many students spent 10 minutes traveling to school?

2. Can you use this plot to determine how many students spent 15 minutes or more traveling to school? Explain why or why not.

3. How many students are there in this class? Describe how you determined your answer.

4. What is the typical time it took for these students to travel to school? Describe how you determined your answer.

Answers

Applications

1. 1 student

2. yes; You can see the individual values by looking at the leaves. There is one student whose travel time is 15 minutes, and there are 16 students whose travel times are more than 15 minutes, for a total of 17 students.

3. 26; You can count the number of leaves on the stem plot. Each value represents one person.

4. Answers will vary. Students may mention the intervals in which the data cluster; most of the students are in the intervals of 0–9 minutes, 10–19 minutes, and 20–29 minutes. Students may find the median, which is 18. They may offer other alternatives as well; however, they must provide clear reasoning for their responses.

In 5–8, use the table below. This table shows the ages, heights, and foot lengths for a group of students.

Students Ages, Heights, and Foot Lengths

Age (months)	Height (cm)	Foot length (cm)
76	126	24
73	117	24
68	112	17
78	123	22
81	117	20
82	122	23
80	130	22
90	127	21
101	127	21
99	124	21
103	130	20
101	134	21
145	172	32
146	163	27
144	158	25
148	164	26
140	152	22
114	135	20
108	135	22
105	147	22
113	138	22
120	141	20
120	146	24
132	147	23
132	155	21
129	141	22
138	161	28
152	156	30
149	157	27
132	150	25

5. Make a stem-and-leaf plot that shows the ages in months of the students, starting with the stem shown here. Notice that the first value in the stem is 6, since there are no values less than 60 months.

```
 6 |
 7 |
 8 |
 9 |
10 |
11 |
12 |
13 |
14 |
15 |
```

6. What ages, in years, does the interval of 80–89 months represent? Explain your answer.

7. What is the median age of these students? Explain how you determined this age.

Connections

8. a. Make a stem plot that shows the heights in centimeters of the students.

   b. Make a line plot of the heights in centimeters of the students.

   c. Which of these plots seems the most appropriate for the data? Why?

   d. Would a bar graph be a good way to show this data? Why or why not?

Extensions

In 9 and 10, use the jump-rope data from Ms. Rich's and Mr. Kocik's classes, which are shown on the next page.

---

5. See below left.

6. $6\frac{1}{2}$ to about $7\frac{1}{2}$ years; Divide the number of months by 12 to convert to years.

7. $113\frac{1}{2}$ months (about $9\frac{1}{2}$ years); There are 30 data values, so the median is the value halfway between the fifteenth and sixteenth values (113 and 114).

Connections

8a. See page 41f.

8b. See page 41f.

8c. The line plot is so spread out that it does not tell us much about the data. The stem plot groups the data and gives us a better picture of the data.

8d. A bar graph would not be any more interest-ing than the line plot. It would be too spread out to tell us much.

5.

Ages in Months

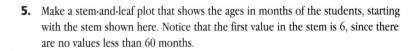

```
 6 | 8
 7 | 3  6  8
 8 | 0  1  2
 9 | 0  9
10 | 1  1  3  5  8
11 | 3  4
12 | 0  0  9
13 | 2  2  2  8
14 | 0  4  5  6  8  9
15 | 2
```

Extensions

9. See page 41g.

10. See page 41h.

Number of Jumps

Mrs. R's Class Data		Mr. K's Class Data	
boy	5	boy	1
boy	35	boy	30
girl	91	boy	28
boy	62	boy	10
girl	96	girl	27
girl	23	girl	102
boy	16	boy	47
boy	1	boy	8
boy	8	girl	160
boy	11	girl	23
girl	93	boy	17
girl	27	boy	2
girl	88	girl	68
boy	26	boy	50
boy	7	girl	151
boy	7	boy	60
boy	1	boy	5
boy	40	girl	52
boy	7	girl	4
boy	20	girl	35
girl	20	boy	160
girl	89	boy	1
boy	29	boy	3
boy	11	boy	8
boy	113	girl	48
boy	33	boy	42
girl	45	boy	33
girl	80	girl	300
		girl	104
		girl	53

9. Make a back-to-back stem-and-leaf plot that compares the girls in Ms. Rich's class with the girls in Mr. Kocik's class *or* the boys in Ms. Rich's class with the boys in Mr. Kocik's class. Did one class of girls (or boys) do better than the other class of girls (or boys)? Explain your reasoning.

10. Make a back-to-back stem-and-leaf plot that compares the girls in both classes with the boys in both classes. Did the girls do better in this activity than the boys? Explain your reasoning.

Mathematical Reflections

In this investigation, you learned how to make stem-and-leaf plots as a way to group a set of data so you can inspect its shape. You looked at two different situations: how long it takes for students to travel to school and how many times students can jump rope. These questions will help you summarize what you have learned:

1 Describe how to locate the median and range using a stem plot.

2 Numerical data can be displayed using more than one kind of graph. How do you decide when to use a line plot, a bar graph, or a stem-and-leaf plot?

3 Some data you gather will be categorical data. Can categorical data be displayed using line plots, bar graphs, or stem-and-leaf plots? Explain your reasoning.

Think about your answers to these questions, discuss your ideas with other students and your teacher, and then write a summary of your findings in your journal.

Think about the survey you will be developing to gather information about middle-school students. What kinds of questions can you ask that might involve using a stem-and-leaf plot to display the data?

Tips for the Linguistically Diverse Classroom

Original Rebus The Original Rebus technique is described in detail in *Getting to Know CMP*. Students make a copy of the text before it is discussed. During discussion, they generate their own rebuses for words they do not understand as the words are made comprehensible through pictures, objects, or demonstrations. Example: Question 1—key words for which students may make rebuses are *median* (a thick vertical line with $\frac{1}{2}$ written on each side), *range* (2–49, 1–63, 5–74), and *stem plot* (a section of a stem plot).

Possible Answers

1. With a stem plot arranged so that all leaves are in ascending order, the range is found by identifying the lowest and the highest values in the data set. The median may be found by first finding half of the number of data values. For example, if there are 46 data values, the median lies between the twenty-third and twenty-fourth values. If there are 45 data values, the median is the twenty-third value. As long as you count consecutively from the smallest value or the largest value, you may count from either end of the data displayed in the stem plot to locate the median.

2. A stem plot is more useful than a line plot or bar graph for data that are spread out. For this type of data, grouping by intervals allows you to see patterns in the data. Line plots are quickly constructed graphs that can be used to "sketch" a data set. If there are a large number of data items, the bar graph is a more useful tool because its vertical scale is adjustable.

3. You can make a graph that resembles a "line" plot using categorical data; the "line" of this type of graph uses categories below it to mark the tally columns. Categorical data can be displayed using a bar graph with categories on the horizontal axis. Categorical data cannot be displayed in a stem plot.

TEACHING THE INVESTIGATION

3.1 • Traveling to School

Launch

To engage students in the context of the problem and the data set, refer them to pages 30 and 31 of the student edition.

> Look at the table of data. What are the three questions that the students asked?

The three questions the students may have asked are, "How long does it take for you to travel to school?" "How far do you travel to get to school?" "How do you get to school?"

> Describe how you think they collected the data to answer each of these questions.

Make sure you review with students how the data are recorded for distance (that is, in decimal form in multiples of a quarter mile, or 0.25). Students will become more comfortable with the data once they review its format.

Refer students to the third question in the "Think about this!" box on page 30. Help them think about what they would have to do to make a line plot to display this data. You may even want to go through the process of trying to create a line plot. Fairly quickly, students will begin to see that the times these students take to travel to school are quite spread out, with a range of 5 to 60 minutes. It would be difficult to make a line plot numbered 5 minutes, 6 minutes, 7 minutes, and so on, to show exact times and—because the data are not really clustered by individual times—it would be difficult to see any patterns in the data. A strategy in which the data are grouped is needed to help us look for patterns in the data.

When students are familiar with the data, introduce the problem of making a stem-and-leaf plot to represent the data.

Explore

The student edition outlines how to develop stem-and-leaf plots. However, we do not recommend that you have students read through this process on their own. Instead, we encourage you to present this problem as a class exploration led by you. Students can consult the student edition for reference at a later time. Here is one way you may proceed.

> Let's work together to build a stem-and-leaf plot. When we look at our data, we see that the range of travel times is 5 minutes to 60 minutes. We can use this information to set up a graph that has a "stem" and several "leaves."

> The "leaves" are the units digits of the data values. The other digits in the data values form the stem. In this case, the stem is made up of the tens digits of the travel times.

For example, if a student takes 45 minutes to get to school, what is the tens digit? (*4*)

What about 15 minutes? (*1*)

What about 5 minutes? (*0*)

What is the highest tens digit we need to show? (*6*)

We show the tens digits as a "stem" of numbers:

```
0 |
1 |
2 |
3 |
4 |
5 |
6 |
```

Next, we begin to add "leaves" to the stem by placing each ones digit next to its tens digits. The first student has a travel time of 60 minutes. We show the ones digit (the 0) as a leaf, like this:

```
0 |
1 |
2 |
3 |
4 |
5 |
6 | 0
```

The next travel times are 15 minutes and 30 minutes. How should I add these to the graph?

Add to the stem plot:

```
0 |
1 | 5
2 |
3 | 0
4 |
5 |
6 | 0
```

The next few travel times are 15 minutes, 15 minutes, 35 minutes, 15 minutes, 22 minutes, 25 minutes, and 20 minutes. Watch how I add these values to the graph.

```
0 |
1 | 5 5 5 5
2 | 2 5 0
3 | 0 5
4 |
5 |
6 | 0
```

I would like you to work with a partner to copy this stem-and-leaf plot and to add the remaining leaves.

Below is the completed stem-and-leaf plot. Notice that the leaves are not in ascending order; they are recorded as they occur in the data list.

```
0 | 8 8 5 5 5 6
1 | 5 5 5 5 9 5 5 7 5 0 5 0 5 1 7 0 0
2 | 2 5 0 5 0 0 0 0 0 1
3 | 0 5 0 0 5
4 |
5 | 0
6 | 0
```

Work with students to rearrange these leaves so they are in order. Transparency 3.1A shows the stem-and-leaf plot before and after arranging the data in ascending order. Help students to add a title to the stem plot and a key for interpreting the plot.

Travel Times to School (minutes)

```
0 | 5 5 5 6 8 8
1 | 0 0 0 0 1 5 5 5 5 5 5 5 5 7 7 9
2 | 0 0 0 0 0 0 1 2 5 5
3 | 0 0 0 5 5
4 |
5 | 0
6 | 0
```

Key

2 | 5 means 25 minutes.

Much of this problem's exploration phase focuses on developing an understanding of the stem-and-leaf plot. The first step in this understanding is making a stem plot. Following this, students need to develop a better understanding of the idea that the data are now grouped in intervals and not simply as a set of the same measure (for example, 15 minutes is grouped with other data in the interval of 10–19).

Summarize

Ask students questions that focus on reading the stem plot and on identifying intervals.

> Look at the 1 stem on the stem-and-leaf plot. What is the shortest travel time shown for this stem? (*10 minutes*)
>
> What is the longest travel time shown for this stem? (*19 minutes*)
>
> What possible travel times are not shown for this stem? (*12 minutes, 13 minutes, 14 minutes, 16 minutes, and 18 minutes*)
>
> We say that the interval of possible times for the 1 stem is 10 to 19 minutes. What is the interval of possible times for the 0 stem? (*0–9 minutes*)
>
> What is the interval of possible times for the 2 stem? (*20–29 minutes*)
>
> What is the interval of possible times for the 3 stem? (*30–39 minutes*)

The questions in this problem guide students to "read the data" and to "read between the data." Before you leave Problem 3.1, spend some time working with them to "read beyond the data."

> How can we describe the shape of the data when it is grouped by tens?

In the Wisconsin class data, most of the data cluster in one area of the stem (between 0 minutes and 35 minutes). There are two outliers: 50 minutes and 60 minutes. Both these students take the bus and probably get on at the beginning of the bus route, since the distance they live from the school is greater than most of the other students.

> Using the mode probably won't tell us too much about the data with this graph. Why do you think this is so?

The mode is the value in the data that occurs most frequently; in these data, the mode is 15 minutes. However, when we look at the stem plot, we are more interested in which interval or intervals contain the most values. This is not the mode, but it may be described as the *modal interval.* In this data, when the data are grouped by tens, it is the interval 10–19.

> How would we find the median for this set of data?

There are 40 measures in this data set. The median is the number halfway between the twentieth and twenty-first values. The second stem plot orders the data, so we can count from either end and locate the twentieth and twenty-first values (15 minutes and 17 minutes). Thus the median is 16 minutes.

As part of the follow-up, you will work with your students to collect data. Specifically, students collect data about the time it takes for them to travel to school, but you may choose to collect data for distance and mode of travel as well. Help your students to develop procedures for finding out how long it takes them to travel to school and, if appropriate, how far they travel to get to school.

To determine how long it takes to get to school, each student can time his or her trip on one morning (or for three or four mornings, and use the median time). To determine the distance, you can post a map that shows the school's location and the surrounding area.

After collecting the data, ask students how they could organize the data to help people make sense of the numbers.

> Would a line plot be a good way to display the data? Why or why not?

> Would a stem plot be a good way to display the data? Why or why not?

After analyzing the data, students describe the typical time it takes them to travel to school. You may also want to have them compare their findings to those of the Wisconsin class.

> What is typical about the time required to travel to school for students in our class?

Students may want to use the median, modal interval(s), or the range to investigate this question. You can assist their discussion by asking further:

> If someone new joined our class today, what would you predict about his or her travel time to school?

3.2 • Jumping Rope

Launch

You may choose to investigate this problem using the data provided, or you may want to help your students conduct their own jump-rope activity and collect their own data. If your class conducts the activity, you will need to develop procedures for collecting the data. (Be aware that collecting these data is time-consuming!) You might ask the physical education teacher to help your students collect the data during their physical education class.

You may want to explore the problem using the data presented in the student edition and then extend the exploration phase to include your students' data, making comparisons where appropriate.

Present the problem by using Transparency 3.2 or by referring students to page 35 in the student edition. Work with your students to make sure they can read the back-to-back stem plot before they begin to work on the problem. One way to do this is to cover the left side of the stem plot and ask students what information is shown on just the right side. Then cover the right side, and have students discuss how the data on the left side are read (when the stem is on the right). Finally, you can show both sets of data together, discussing how this arrangement lets you make comparisons between data sets.

Explore

Once students are comfortable with the data display, they can focus on the question posed in Problem 3.2. Students should work in small groups on the problem and follow-up.

Summarize

Hold a class discussion about Problem 3.2. The process of comparison may be difficult. Students may wonder how they can compare data sets that contain different numbers of data items. They may not immediately think about locating the medians and ranges of the two classes' data, yet these are precisely the tools that can help them make comparative statements.

Additional Answers

ACE Questions

Applications

8a.

Heights in Centimeters

11	2 7 7
12	2 3 4 6 7 7
13	0 0 4 5 5 8
14	1 1 6 7 7
15	0 2 5 6 7 8
16	1 3 4
17	2

8b.

Student Heights

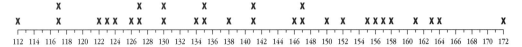

Height of students (cm)

9. The graph below uses *G* and *B* in place of actual numbers of jumps (refer to earlier stem plot for numbers), giving girls' and boys' data on the same plot. Students will have made separate plots, but you may use this summary graph as a way to show how we can modify stem plots to give different information.

Numbers of Jumps

Mr. K's class		Ms. R's class
B B B G B B B B B	0	B B B B B B B
B B	1	B B B
B G G	2	G B G B G B
G B B	3	B B
G B G	4	B G
G G B	5	
G B	6	B
	7	
	8	G G G
	9	G G G
G G	10	
	11	B
	12	
	13	
	14	
G	15	
B G	16	
	17	
	18	
	19	
	20	
	21	
	22	
	23	
	24	
	25	
	26	
	27	
	28	
	29	
G	30	

Generally speaking, girls performed similarly in both classes, as did boys. We can see a couple of outliers for boys—one in each class. One girl in Mr. Kocik's class jumped rope more times than anyone else in either class.

10. One way to answer this question is to show all the girls' data on one side of the stem plot and all the boys' data on the other side, as shown below.

Numbers of Jumps

Girls	Stem	Boys
4	0	1 1 1 1 2 3 5 5 7 7 7 8 8 8
	1	0 1 1 1 6 7
7 7 3 3 0	2	0 6 8 9 6 0
5	3	0 3 3 5
8 5 2	4	0 2 7
3 2	5	0
8	6	0 2
	7	
9 8 0	8	
6 3 1	9	
4 2	10	
	11	3
	12	
	13	
	14	
1	15	
0	16	0
	17	
	18	
	19	
	20	
	21	
	22	
	23	
	24	
	25	
	26	
	27	
	28	
	29	
0	30	

The boys' data clusters at the lower end of the stem plot. The girls' data is spread out with more of the data showing larger numbers of jumps.

Coordinate Graphs

Mathematical and Problem-Solving Goals

- **To implement the process of statistical investigation to answer questions**

- **To review the process of measuring length, time, and distance**

- **To analyze data by using coordinate graphs to explore relationships among variables**

- **To explore intervals for scaling the vertical axis (y-axis) and the horizontal axis (x-axis)**

Student Pages   42–52

Teaching the Investigation   52a–52g

In this investigation, students use coordinate graphs to examine relationships between two measures. In Problem 4.1, Relating Height to Arm Span, they measure heights and arm spans of all students in the class and make a coordinate graph to explore whether there is a relationship between the two measures.

In Problem 4.2, Relating Travel Time to Distance, students return to a data set used in Investigation 3 that displays the time required to travel to school for several students. They use a coordinate graph of this data to determine whether there is a relationship between the times the students spend traveling to school and the distances between their homes and school.

Materials

For students

- Labsheets 4.2 and 4.ACE (1 each per student)
- Grid paper (provided as a blackline master)
- Yardsticks, meter sticks, or tape measures
- Chart paper with a 1-inch grid (optional; available at an office-supply store)
- String (optional)

For the teacher

- Transparencies 4.1A, 4.1B, and 4.2 (optional)
- Chart paper with a 1-inch grid (optional; available at an office-supply store)
- Colored stick-on dots (optional)

4.1

Relating Height to Arm Span

Launch

- Help students to understand coordinate graphs by analyzing the example in the student edition.

- Discuss ways of collecting class data on height and arm span measurements, and collect the data.

- Discuss with students how the class's data might be displayed, and analyze the strengths and weaknesses of various representations.

Explore

- Have students make coordinate graphs to represent the data.

Summarize

- Talk about choosing a scale for each axis. Review questions students should ask themselves when deciding how to label the axes.

- Talk about the follow-up questions, helping students to understand what the different areas of the graph represent.

Assignment Choices

ACE questions 1–3 (2 requires grid paper) and unassigned choices from earlier problems

Coordinate Graphs

In the first three investigations, you worked with one measure at a time. For example, you looked at the number of letters in students' names, the number of times students jumped rope, or the numbers of pets students had. Although you can find out some interesting things about one set of data, it is often interesting to look at how two sets of data are related to each other.

4.1 Relating Height to Arm Span

If you look around at your classmates, you might guess that taller people have wider arm spans. But is there *really* any relationship between a person's height and his or her arm span? The best way to find out more about this question is to collect some data.

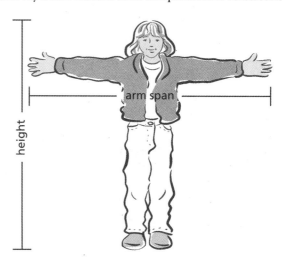

Here are data on height and arm span (measured from fingertip to fingertip) that one class collected.

Height and Arm Span Measurements

Initials	Height (inches)	Arm span (inches)
NY	63	60
JJ	69	67
CM	73	75
PL	77	77
BP	64	65
AS	67	64
KR	57	57

One way to show data about two different measures (such as height and arm span) at the same time is to make a **coordinate graph.** Each point on a coordinate graph represents two measures for one person or thing. On a coordinate graph, the horizontal axis, or **x-axis,** represents one measure. The vertical axis, or **y-axis,** represents a second measure. For example, the graph below shows data for height along the x-axis and data for arm span along the y-axis. Each point on this graph indicates both the height and the arm span for one student.

Height and Arm Span Measurements

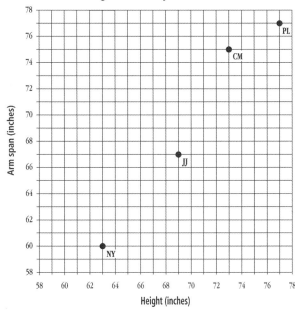

Study the table of data and the coordinate graph. Four points have already been plotted and labeled with the students' initials. We write the location for each point like this:

- NY is located at point (63, 60)
- JJ is located at point (69, 67)
- CM is located at point (73, 75)
- PL is located at point (77, 77)

Working with a partner, determine how to locate points on this graph. Where would you place the points and initials for the remaining three people? Why do the axes start at (58, 58)? What would the graph look like if the axes started at (0, 0)?

Problem 4.1

Think about this question: If you know the measure of a person's arm span, do you know anything about his or her height?

To help you answer this question, you will need to collect some data. With your class, collect the height and arm span of each person in your class. Make a coordinate graph of your data. Then, use your graph to answer the question above.

■ **Problem 4.1 Follow-Up**

Draw a diagonal line through the points on the graph where the measures for arm span and height are the same.

1. How many of your classmates' data are on this line? What is true about arm span compared to height for the points on this line?

2. What is true about arm span compared to height for the points *below* the line you drew?

3. What is true about arm span compared to height for the points *above* the line you drew?

Answer to Problem 4.1

If you know the measure of a person's arm span, you can reasonably guess that the person's height is approximately the same.

Answers to Problem 4.1 Follow-Up

1. Answers will vary. The line is the graph of $y = x$. Students do not need to understand this equation now. Instead, focus on understanding what students know about the data that are located on, above, or below the line. They should understand that points on the line represent students whose heights are equal to their arm spans.

2. Points below the line represent students whose heights are greater than their arm spans.

3. Points above the line represent students whose arm spans are greater than their heights.

4.2 Relating Travel Time to Distance

In Investigation 3, you made stem plots to show data about travel times to school. Using a coordinate graph, you can show both travel time and distance from home to school. For example, we can show the travel time and distance to school for the students in Problem 3.1 on a coordinate graph:

Times and Distances to School

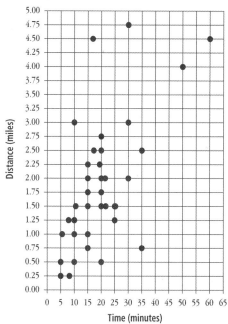

Problem 4.2

Study the graph above, which was made using the data from Problem 3.1.

A. Look back at the data on page 31. On Labsheet 4.2, locate and label with initials the points for the first five students in the table.

B. If you know how long it takes a particular student to travel to school, can you know anything about that student's distance from school? Use the graph to help you answer this question. Write a justification for your answer.

Answers to Problem 4.2

A. See page 52f.

B. The graph seems to go up in distance as time increases. So students who live further away from school generally spend more time getting to school. But you must also consider how students get to school (e.g., walking, riding a bike, by car, or by bus).

At a Glance

Launch

■ With students, study the coordinate graph that displays the data from Problem 3.1.

Explore

■ Have students work on Problem 4.2.

Summarize

■ Invite several pairs to share their answers to part B.

■ Ask questions to clarify how to read the graph.

Assignment Choice

ACE question 5 (5 requires Labsheet 4.ACE) and unassigned choices from earlier problems

■ **Problem 4.2 Follow-Up**

1. Locate the points at (17, 4.50) and (60, 4.50) on the coordinate graph on Labsheet 4.2. What can you tell about the students these points represent?

2. Locate the points (30, 2.00), (30, 3.00), and (30, 4.75). What can you tell about the students these points represent?

3. What would the graph look like if the same scale were used for both axes?

Answers to Problem 4.2 Follow-Up

1. These students live the same distance from school, but one takes an hour to get to school, and the other takes only 17 minutes. Maybe one goes by car and one by bus. Or perhaps, both go by bus, but one is picked up earlier on the route.

2. These three students take the same time to get to school, but live very different distances from school. Maybe one walks, one bikes, and one comes by car.

3. The range of times is 5 minutes to 60 minutes, but the range of distances is only $\frac{1}{4}$ to $4\frac{3}{4}$ miles. If we used the scale from the time axis on the distance axis, all the points would be on top of each other. If we used the scale from the distance axis on the time axis, the graph would not fit on a piece of paper.

Applications • Connections • Extensions

As you work on these ACE questions, use your calculator whenever you need it.

Applications

In 1 and 2, use this table of data, which shows the ages, heights, and foot lengths for 30 students.

Student Age, Height, and Foot Length

Age (months)	Height (cm)	Foot length (cm)
76	126	24
73	117	24
68	112	17
78	123	22
81	117	20
82	122	23
80	130	22
90	127	21
101	127	21
99	124	21
103	130	20
101	134	21
145	172	32
146	163	27
144	158	25
148	164	26
140	152	22
114	135	20
108	135	22
105	147	22
113	138	22
120	141	20
120	146	24
132	147	23
132	155	21
129	141	22
138	161	28
152	156	30
149	157	27
132	150	25

Answers

Applications

1a. See below right.

1b. You cannot tell whether the youngest student is also the shortest from the stem plot because data for age is not displayed. You can use the table because you can match height and age; however, you cannot do this quickly, because the data are not ordered.

2a. See page 52f.

2b–d. See page 52g.

3a. The graph indicates that, in general, taller people have longer foot lengths. However, knowing a person's foot length will not definitively tell you that person's height.

3b. The median height is approximately $139\frac{1}{2}$ cm (the value halfway between 138 cm and 141 cm). The median foot length is 22 cm. Dividing 139 by 22, we see that the median height is a little more than 6 times the median foot length.

3c. Answers will vary. Height is generally about 6 to $6\frac{1}{2}$ times foot length.

3d. The answers to b and c show that a person's height is generally 6 to $6\frac{1}{2}$ times his or her foot length, so we can use foot length to estimate height (however, we cannot know the exact height for certain).

3e. The data would be bunched in the upper-right corner of the graph.

---

1. **a.** Make a stem-and-leaf plot showing the heights of these students.

b. Can you determine from your stem plot whether the youngest student is also the shortest? Can you determine this from the table? Explain why or why not.

2. **a.** On a piece of grid paper, make a coordinate graph showing each person's age (in months) on the horizontal axis and height (in centimeters) on the vertical axis. To help you choose a scale for each axis, look at the range of values you have to locate on the graph. What are the smallest and largest age values? What are the smallest and largest height values?

b. Can you determine from your coordinate graph whether the youngest student is also the shortest student? Explain your reasoning.

c. Using information from your coordinate graph, describe what happens to students' heights as students get older.

d. We know people eventually stop growing. When does this happen? How would this affect the graph?

3. The coordinate graph on the next page displays height and foot length for 28 students. Notice that the *x*-axis is scaled in intervals of 5 centimeters and the *y*-axis is scaled in intervals of 1 centimeter.

a. One student said that if you know a person's foot length, you can tell what that person's height is. Do you think she is right? Explain your reasoning.

b. Determine the median height and the median foot length. Compare the median height with the median foot length. The median height is about how many times as large as the median foot length?

c. Measure the length of your foot and your height in centimeters. Your height is about how many times as large as your foot length?

d. Look at your responses to b and c. How can you use this information to decide whether the student's comment in a is correct?

e. What would the graph look like if each axis started at 0?

1a. **Heights of Students**

11	2 7 7
12	2 3 4 6 7 7
13	0 0 4 5 5 8
14	1 1 6 7 7
15	0 2 5 6 7 8
16	1 3 4
17	2

Heights and Foot Lengths

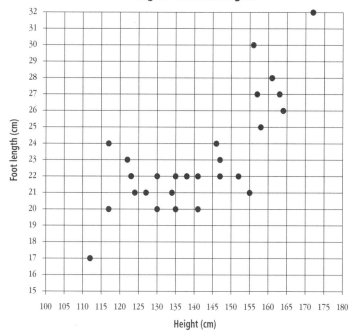

Connections

4. See below right for graph. Patterns students may see: The horizontal line through two factors shows prime numbers. The only number with one factor is 1. The numbers that are products of two different primes (6, 10, 14, and 15) have four factors.

Connections

4. Make a coordinate graph that shows the numbers from 1 to 20 on the horizontal axis and the number of factors of each of these numbers on the vertical axis. What patterns do you see in your graph? Explain each pattern.

Factors of Whole Numbers

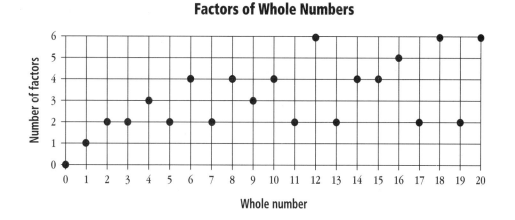

Extensions

5a. The actual counts have a range of 309 to 607 seeds. The actual counts fall within a small range compared to the guesses. The median is $459\frac{1}{2}$ seeds (halfway between 455 and 462).

5b. The guesses have a range of 200 to 2000 seeds. The guesses are much more spread out than the actual counts. The median is $642\frac{1}{2}$ seeds (halfway between 630 and 655).

5c. See page 52g.

5d. These points represent guesses that are very close or equal to the actual counts.

5e. These points represent guesses that are larger than the actual counts.

5f. These points represent guesses that are smaller than the actual counts.

5g. In general, the guesses are larger than the actual counts. The median for the actual counts ($459\frac{1}{2}$) is much smaller than the median for the guesses ($642\frac{1}{2}$), but the amazing thing is the difference in the ranges. The difference between the smallest and largest guesses is 1800 seeds, while the difference between the smallest and largest actual counts is only 298 seeds. The range for the guesses spans six times as many values.

5h. Possible answer: You could change the scale on the horizontal axis to go from 0 to 750.

Extensions

5. A group of students challenged each other to see who could come the closest to guessing the number of seeds in his or her pumpkin. In October, they guessed the number of seeds in each of their pumpkins. In November, they opened their pumpkins and counted the seeds. They compared their guesses with their actual counts by displaying their data on a coordinate graph. The data and graph are shown on the next page.

a. Describe what you notice about how spread out the actual counts are. What are the median and the range of the actual counts?

b. Describe what you notice about how spread out the guesses are. What are the median and the range of the guesses?

c. On Labsheet 4.ACE, draw a diagonal line on the graph to connect the points (0, 0), (250, 250), (500, 500), all the way to (2500, 2500).

d. What is true about the guesses compared to the actual counts for points on or near the line you drew?

e. What is true about the guesses compared to the actual counts for points above the line?

f. What is true about the guesses compared to the actual counts for points below the line?

g. In general, did the students make good guesses about the numbers of seeds in their pumpkins? Use what you know about the median and range of the actual counts and the guesses as well as other information from the graph to explain your reasoning.

h. The scales on the axes are the same, but the data is very bunched together. How would you change the scale to better show the data points?

Numbers of Seeds in Pumpkins

Guess	Actual
630	309
621	446
801	381
720	505
1900	387
1423	336
621	325
1200	365
622	410
1000	492
1200	607
1458	498
350	523
621	467
759	423
900	479
500	512
521	606
564	494
655	441
722	455
202	553
621	367
300	442
200	507
556	462
604	384
2000	545
1200	354
766	568
624	506
680	486
605	408
1100	387

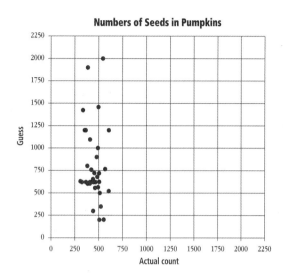

Numbers of Seeds in Pumpkins

Possible Answers

1. You assign the first measure for each pair to the *x*-axis, and the second measure to the *y*-axis. Then, you consider the range of the data as you set up the scale of each axis.

2. To locate a point, give the location along the horizontal axis (the *x*-axis) followed by the location along the vertical axis (the *y*-axis).

3. The points will tend to rise as you look from left to right.

4. A coordinate graph and a line plot both show numerical data. A line plot shows only one measure or count. A coordinate graph shows two measures or counts for the same object (or person). You could start with a coordinate graph and make a line plot for the values represented along the *x*-axis and a different line plot for values represented along the *y*-axis, but the line plots would not show which pairs of values go together.

In this investigation, you have learned how to make and read coordinate graphs. Coordinate graphs let you examine two sets of data at once so you can look for relationships between pairs of data. You looked at two different situations: how arm span relates to height, and how travel time relates to distance traveled. These questions will help you summarize what you have learned:

1 When you make a coordinate graph of data pairs, what do you consider when deciding what scale to use on each axis?

2 How do you locate a point on a coordinate graph?

3 If you make a coordinate graph of variables such as arm span and height, where the values of one measure get larger as the values of the other measure get larger, what will the pattern of points on the graph look like?

4 How are a coordinate graph and a line plot alike? How are they different?

Think about your answers to these questions, discuss your ideas with other students and your teacher, and then write a summary of your findings in your journal.

Think about how what you have learned about coordinate graphs might help you with your project. What kinds of questions can you ask to help you answer the question, "Is anyone typical?" that might involve using a coordinate graph to display the data?

Tips for the Linguistically Diverse Classroom

Diagram Code The Diagram Code technique is described in detail in *Getting to Know CMP*. Students use a minimal number of words and drawings or diagrams to respond to questions that require writing. Example: Question 2—A student might answer this question by drawing a set of axes with an arrow pointing to 5 on the *x*-axis and a second arrow extending vertically from the end of the first arrow to a height of 10 on the *y*-axis. At the point of the second arrow, the student might make a point and label it (5, 10).

TEACHING THE INVESTIGATION

4.1 • Relating Height to Arm Span

Launch

Introduce coordinate graphs by using the data and graph shown on page 43 of the student edition. You can use Transparency 4.1A to help you explain coordinate graphs. If you do not have an overhead projector, you could use a large sheet of chart paper with a 1-inch grid, and label the axes as they appear in the student edition. Various students can locate the points of the sample data set by placing colored dots (with initials of each person in the data set written on them) on the grid. Focus students on the axes, and discuss how they are labeled and scaled.

For the Teacher

We suggest that the dots have initials on them to help students understand how to locate points. Be careful *not* to send the message that this is necessary when constructing coordinate graphs. On a graph of the entire class's data, initials would take up too much space and make it hard to read the graph and look for patterns.

When you feel students understand the example, talk about how to collect the class's data. Discuss strategies for how students can measure their heights and arm spans to the nearest inch (or centimeter). For example, students can mark their heights and arm spans on pieces of string and then measure the string with a yardstick or meter stick. Or they can tape yardsticks or meter sticks to the wall and stand against them. Remind students that you are paying attention to accuracy and that they need to maintain consistency in their data-collection techniques with regard to such things as students' shoe heels.

> Working with your partner, measure each other's height and arm span in inches (or centimeters). We will record our data in a central location so we can all see and use it.

For the Teacher

You may want to have students remove their shoes before measuring their height. We have found that when students leave their shoes on, their height measurements are less accurate; that is, their heights are disproportionately greater than their arm spans.

If students have trouble measuring accurately, affix a measuring tape to a wall vertically so students can stand against it, and affix another horizontally to a wall at shoulder height to help students measure their arm spans.

Use the board or a large sheet of chart paper to display the data. Have students write their names and record their two measures so the data may be seen by all students. You can organize your table of measures like the sample table in the student edition.

Once the data have been gathered and students are ready to work on organizing and interpreting it, bring the class back together to revisit the problem statement.

Explain again that we are interested in how the two variables are, or seem to be, related. A coordinate graph is an ideal representation for analyzing the question. This is in fact the strength of coordinate graphs: showing how two variables are related.

> How might we organize and display the data in a graph to help us answer the question?

In their study of the process of data investigation, your students have already developed many ways to represent data. It is important that they see why the techniques they have already mastered are not as effective as a coordinate graph for displaying the data for this problem. You may decide to propose alternative representations for the data to help your students understand why these representations do not give the best display. Help your students begin to see the different strengths and weaknesses of the various forms of graphical representation.

You could propose using a double bar graph, with two bars for each student—one showing arm span and the other showing height. Here's an example of specific data and a double bar graph.

Table of Measures

Initials	Height (inches)	Arm span (inches)
WE	63	60
AS	64	65
JF	64	65
AD	67	64
LP	68	68
JJ	69	67
GL	69	71
NY	70	68
BP	70	65
SH	70	73
CM	70	69
RS	71	70
KR	72	72
PL	73	75
JD	77	77

Height and Arm Span Measures

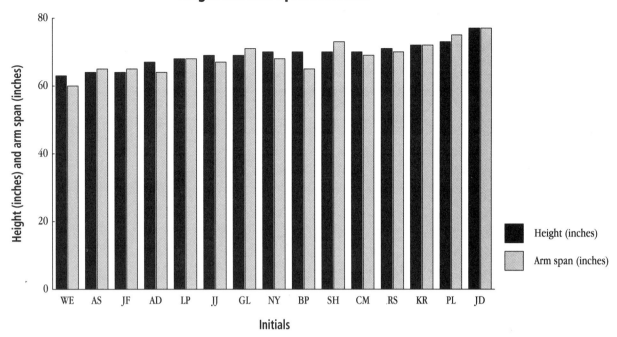

Help students see that the double bar graph representation is tedious to make because it requires many sets of bars. Since the data in the double bar graph are not grouped or ordered in any way, it is difficult to determine how to order the bars. However, you can see that the bars in each pair are about the same length, indicating a relationship between height and arm span.

You might also propose making a back-to-back stem plot with height on one side and arm span on the other.

Height and Arm Span Measures

Height		Arm span
	5	
9 9 8 7 4 4 3	6	0 4 5 5 5 7 8 8 9
7 3 2 1 0 0 0 0	7	0 1 2 3 5 7
	8	

This plot shows that the shape of both sets of data are similar, but it does not show which height is paired with which arm span. This makes it hard to look for a relationship.

Explore

Ask your students to make a coordinate graph to display their class's data. Discuss how they might scale and label the axes, including the idea that the scale must be consistent with the measures. Here are some questions that students should consider (you may want to write these on the board):

■ Which measure should go on the horizontal axis and which should go on the vertical axis? (Either is correct, but usually we put the measure we want to use to predict the other measure on the horizontal axis.)

- What is the range of measures that you need to show on the horizontal axis? The vertical axis?

- What would your graph look like if you just started each axis at 0?

- What labeling scheme will show all of the data in the space you have for the graph?

- Will the scale you have chosen spread the data out too much or bunch it up so that it is hard to see patterns?

When you feel your students have a good idea about what scales make sense, allow them to make the graph and to answer the questions in the follow-up.

Summarize

Let groups share their graphs and tell how they chose the scale or labeling scheme for each axis. Discuss which graphs seem to show the data best and why.

Discuss the answers to the follow-up questions. Help students understand what the three different regions of the graph represent: the line represents points where height equals arm span; the region above the line represents points where arm span is greater than height, and the region below the line represents points where height is greater than arm span.

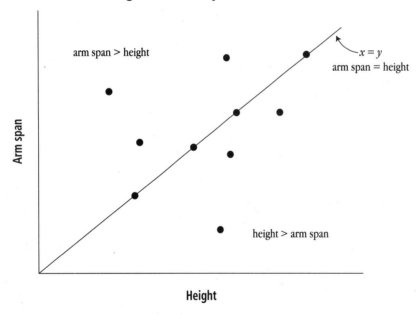

Height and Arm Span Measures

4.2 • Relating Travel Time to Distance

Launch

In this problem, students use the data introduced in Investigation 3.

> In the last problem, you were introduced to coordinate graphs. These are interesting graphs because they allow us to look at two sets of data at the same time, so we can find relationships between them.

Refer students to the data on page 45 of the student edition or display Transparency 4.2. Spend time discussing the scales for each axis so students understand how the axes relate to the data in the table. You might choose to recreate the graph on a large sheet of chart paper with a 1-inch grid. This will allow students to better understand how the graph was created before they spend time discussing what it tells them. You could have various students locate the points by placing colored stick-on dots on the chart-paper grid or by making a dot on the transparency grid with a marker.

Explore

Have students—individually or in pairs—write their justification for their answer to question B about the relationship between the time required to travel to school and distance traveled. Have them work on Problem 4.2 Follow-Up or save these questions for the class summary.

Summarize

Refocus students on the big question.

> If you know how long it takes a particular student to travel to school, can you know anything about that student's distance from school?

Let several students or pairs read their justifications. This may raise an opportunity to ask such questions as the following:

> Look at the axis showing time. How can you tell from the graph the shortest time it took someone to travel to school? The longest time?

> Look at the axis showing distance. How can you tell from the graph the shortest distance it took someone to travel to school? The longest distance?

Discuss the answers to the follow-up questions. These questions help deepen students' understanding of the data displayed on the graph.

Additional Answers

Answers to Problem 4.2

A.

Times and Distances to School

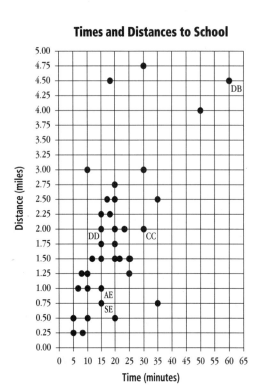

ACE Questions

Applications

2a.

Heights and Ages

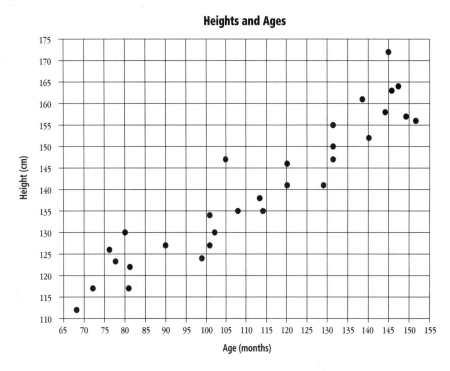

2b. You can locate the youngest student (the furthest to the left on the horizontal axis) and the shortest student (the closest to the bottom on the vertical axis). You can quickly see that the youngest student *is* the shortest student.

2c. In general, older students are taller.

2d. People do eventually stop growing in their late teens or early twenties. The graph would level out for these ages, and we would not see much increase afterward.

Extensions

5c.

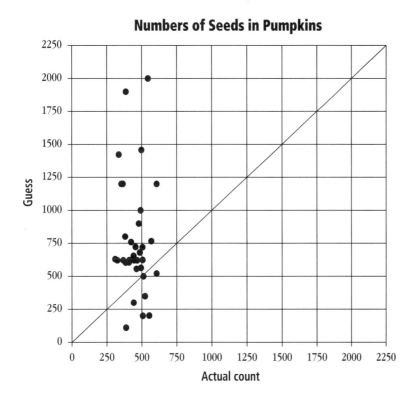

What Do We Mean by *Mean?*

Mathematical and Problem-Solving Goals

- **To understand the mean as a number that "evens out" or "balances" a distribution**

- **To create distributions with a designated mean**

- **To find the mean of a set of data**

- **To use the mean to help describe a set of data**

- **To reason with a model that clarifies the development of the algorithm for finding the mean**

- **To distinguish between the mean, median, and mode as ways to describe what is typical about a set of data**

Student Pages   53–68

Teaching the Investigation   68a–68n

This investigation focuses on the development of the concept of *mean.* The notions of "evening out" and of "balancing" a distribution at a point (the mean) on the horizontal axis are modeled by using interlocking cubes and stick-on notes. These models support the development of the algorithm for finding the mean: adding up all the numbers and dividing by the total number of numbers.

In Problem 5.1, Evening Things Out, students explore different ways to describe the average number of people in eight households. The mean is introduced through a visual model using interlocking cubes. In Problem 5.2, Finding the Mean, students must apply what they have learned in Problem 5.1. Problem 5.3, Data with the Same Mean, links Problems 5.1 and 5.2 and explores the idea that different sets of data may yield the same mean.

In Problem 5.4, Using Your Class's Data, students use a larger data set than they have encountered in the previous problems. This motivates them to develop the algorithm. In Problem 5.5, Watching Movies, students explore what happens to the mean when very small or very large data values are added to a data set.

Research has shown that students can learn the algorithm for finding the mean with relative ease. This investigation seeks to develop an understanding of what the mean represents and what it tells us about a set of data.

Materials

For students

- Stick-on notes
- Interlocking cubes (10 each of 9 different colors per student)
- Large sheets of unlined paper
- Colored pens, pencils, or markers

For the teacher

- Transparencies 5.1A, 5.1B, 5.2, 5.3, 5.4, and 5.5 (optional)

What Do We Mean by *Mean?*

Since the first United States census was conducted in 1790, its primary use has been to find out how many people live in the United States. These data, organized by state, are used to determine how many representatives each state will have in the House of Representatives in the United States Congress.

Many people are interested in the census because it provides useful information about a number of other things, including household size. The term *household,* as used by the United States census, means all the people who occupy a "housing unit" (a house, an apartment or other group of rooms, or a single room like a room in a boarding house).

Remember that an *average* is a value used to describe what is typical about a set of data. An average can be thought of as a "measure of center." The mode and the median are two types of averages you have used quite a bit. The *mode* is the value that occurs most frequently in a set of data. The *median* is the value that divides a set of ordered data in half. This investigation explores a third kind of average, which is called the *mean*.

Tips for the Linguistically Diverse Classroom

Visual Enhancement The Visual Enhancement technique is described in detail in *Getting to Know CMP*. It involves using real objects or pictures to make information comprehensible. Example: While discussing this page, you might show a map of the United States, pictures of Congress and the nation's Capitol, and photos of a variety of types of housing.

Evening Things Out

At a Glance

Launch

- Talk with students about the United States census and the definition of *household*.

- Introduce the problem through a demonstration of how to construct the tower representation of the data.

Explore

- Work on the problem by exploring how to construct the line-plot representation of the data.

Summarize

- Ask students to write a definition of the word *mean*.

5.1 Evening Things Out

Eight students in a middle-school class determined the number of people in their households using the United States census guidelines. Each student then made a cube tower to show the number of people in his or her household.

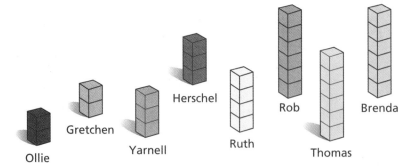

You can easily see from the cube towers that the eight households vary in size. The students wondered what the average number of people is in their households. Their teacher asked them what they might do, using their cube towers, to find the answer to their question.

> **Problem 5.1**
>
> What are some ways to determine the average number of people in these eight households?

■ **Problem 5.1 Follow-Up**

The students had an idea for finding the average number of people in the households. They decided to rearrange the cubes to try to "even out" the number of cubes in each tower. Try this on your own and see what you find for the average number of people in the households, and then read on to see what the students did.

Assignment Choices

ACE question 1 and unassigned choices from earlier problems

Answers to Problem 5.1

Answers will vary. Students may suggest finding the median or mode. Some students may suggest adding up the values and dividing by 8.

First, the students put the towers in order.

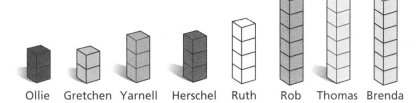

Ollie   Gretchen   Yarnell   Herschel   Ruth   Rob   Thomas   Brenda

The students then moved cubes from one tower to another, making some households bigger than they actually were and making other households smaller than they actually were. When they were finished moving cubes, their towers looked like this:

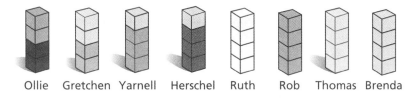

Ollie   Gretchen   Yarnell   Herschel   Ruth   Rob   Thomas   Brenda

Each tower now had four cubes. Notice that the total number of cubes did not change.

Before		After	
Ollie	2 people	Ollie	4 people
Gretchen	2 people	Gretchen	4 people
Yarnell	3 people	Yarnell	4 people
Herschel	3 people	Herschel	4 people
Ruth	4 people	Ruth	4 people
Rob	6 people	Rob	4 people
Thomas	6 people	Thomas	4 people
Brenda	6 people	Brenda	4 people
Total	32 people	Total	32 people

The students determined that the average number of people in a household was 4. The teacher explained that the average the students had found is called the **mean**. The mean number of people in the eight households is 4.

The students decided to look at the data in another way. They used stick-on notes to make a line plot of the data. They used an arrow to show the mean on their line plot.

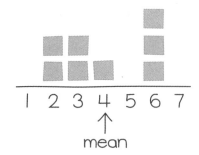

Notice that in the 8 households there is a total of 32 people: 2 + 2 + 3 + 3 + 4 + 6 + 6 + 6 = 32. You can see that the mean is not the middle of the distribution since there are 3 households above the mean and 4 households below the mean.

One student said, "When you even out the cubes, it's like moving all the stick-on notes to the same place on the line plot." The students showed this on their line plot.

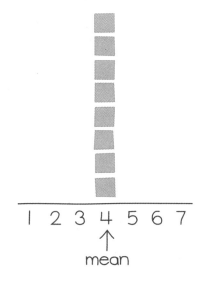

Notice that the total number of people in the households is still 32: 4 + 4 + 4 + 4 + 4 + 4 + 4 + 4 = 32.

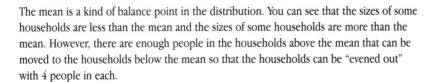

The mean is a kind of balance point in the distribution. You can see that the sizes of some households are less than the mean and the sizes of some households are more than the mean. However, there are enough people in the households above the mean that can be moved to the households below the mean so that the households can be "evened out" with 4 people in each.

5.2 Finding the Mean

The following data show the number of people in the households of eight different students.

Name	Number of people in household
Geoffrey	6
Betty	4
Brendan	3
Oprah	4
Yancey	2
Reilly	3
Tonisha	4
Barker	6

Problem 5.2

A. Make a set of cube towers to show the size of each household.

B. Make a line plot of the data.

C. How many people are there in the eight households altogether? Describe how you determined your answer.

D. What is the mean number of people in the eight households? Describe how you determined your answer.

Problem 5.2 Follow-Up

1. How does the mean for this set of eight students compare to the mean for the eight students in Problem 5.1?

2. How does the median for this set of eight students compare to the median for the eight students in Problem 5.1?

Answers to Problem 5.2

A. See page 67l.

B. See page 67l.

C. 32; Possible answer: count the total number of cubes.

D. 4; Possible answer: If you "even out" the towers, there are four cubes in each tower.

Answer to Problem 5.2 Follow-Up

1. The means are the same.

2. The median household size for these students is 4. The median of the data in Problem 5.1 is 3.5.

At a Glance

Launch

- Distribute the materials for building the two representations.

- Make sure students understand the task.

Explore

- Circulate, helping students with their tower and line-plot models.

Summarize

- In a class discussion, have students share their strategies for constructing the models.

Assignment Choices

ACE question 2 and unassigned choices from earlier problems

Data with the Same Mean

At a Glance

Launch

- With students, examine the line plots for the data in Problems 5.1 and 5.2.

- As a class, develop a data set with the same mean as the first two sets.

Explore

- Have pairs find data sets that meet the criteria of the problems.

- Post one line plot from each pair, and examine the line plots as a class.

Summarize

- Synthesize what students know about making distributions to meet different criteria.

- Have students make cube towers for an example you present.

5.3 **Data with the Same Mean**

The line plots below show the data from Problem 5.1 and Problem 5.2. The data for the two situations look different, but the mean is the same.

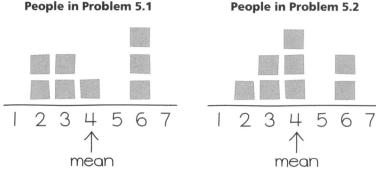

Think about these questions:
- How many households are there in each situation?
- How many total people are there in each situation?
- How are these facts related to the mean of 4 in each case?

Problem 5.3

A. Try to find two more sets of eight households with a mean of 4 people. Use cubes to show each set, and then make line plots that show the information from the cubes.

B. Try to find two different sets of nine households with a mean of 4 people. Use cubes to show each set, and then make line plots to show the information from the cubes.

Problem 5.3 Follow-Up

1. A group of nine students has a mean number of 3 people in their households. Make a line plot showing a data set that fits this description.

2. A group of eight students has a mean number of $3\frac{1}{2}$ people in their households. Make a line plot showing a data set that fits this description.

3. How can the mean be $3\frac{1}{2}$ people when "half" people do not exist?

---

58 Data About Us

Assignment Choices

ACE questions 3–4 and unassigned choices from earlier problems

Answers to Problem 5.3

See page 67l.

Answers to Problem 5.3 Follow-Up

1. See page 67m.

2. See page 67m.

3. The mean is the number obtained by dividing the sum of the data values equally among the households. In question 2, the sum of the data is 28. Since 8 does not divide evenly into this number, the mean is not a whole number. Unlike the mode, the mean does not have to be an actual (or even a possible) value in the data set.

Recall that the term *household* as used by the United States census refers to all people who occupy a "housing unit" (a house, an apartment or other group of rooms, or a single room like a room in a boarding house).

Problem 5.4

A. Using the definition from the United States census, how many people are in your household?

B. Collect household data from everyone in your class, and make a display to show the information.

C. What is the mean number of people in your class's households? Describe how you determined your answer.

▨ Problem 5.4 Follow-Up

One student wrote, "In Problem 5.2, there are 8 households with a total of 32 people. There is a range of 2 to 6 people in the 8 households. The mean number of people in each household is 4. This is the number that tells me how many people each household would have if each household had the same number of people." Write a similar description for your class data about people in your households.

Answer to Problem 5.4

Answers will vary.

Answer to Problem 5.4 Follow-Up

Answers will vary.

Using Your Class's Data

- - - - - - - - - -
At a Glance

Launch

■ Ask questions to help students verbalize what they know about strategies for finding the mean of a data set.

Explore

■ Have groups or pairs collect data on the number of people in each student's household, make a display of the data, and calculate the mean of the data.

■ Have students write a description summarizing what they discovered.

Summarize

■ Synthesize what students know about making distributions to meet different criteria.

■ Have students make cube towers for an example you present.

Assignment Choices

ACE questions 5–7 and unassigned choices from earlier problems

5.5

Watching Movies

A group of middle-school students was asked this question: How many movies did you watch last month? Here are a table and stem plot of the data:

Student	Number of movies
Joel	15
Tonya	16
Rachel	5
Lawrence	18
Meela	3
Leah	6
Beth	7
Mickey	6
Bhavana	3
Josh	11

Movies Watched

```
0 | 3 3 5 6 6 7
1 | 1 5 6 8
2 |
```

Key
1 | 5 means 15 movies.

At a Glance

Launch

■ Talk with students about the different ways the question about the number of movies watched could be interpreted.

Explore

■ Have pairs explore the stem plot and the mean for each part of the problem.

Summarize

■ Hold a class discussion about what students discovered.

■ Ask questions to strengthen students' understanding of the effect of new data values on the stem plot and the mean.

Problem 5.5

A. Look at the table above and complete these statements.

The total number of students is _____.

The total number of movies watched is _____.

The mean number of movies watched is _____.

B. A new value is added for Lucia, who watched 42 movies last month. This value is an outlier. How does the stem plot change when this value is added? What is the new mean? Compare the mean from part A to the mean after this value is added. What do you notice?

C. A new value is added for Tamara, who was home last month with a broken leg. She watched 96 movies. What is the mean of the data now? Compare the means you found in parts A and B with this new mean. What do you notice? Why?

D. Data for eight more students are added:

Tommy	5	Robbie	4
Alexandra	5	Ana	4
Kesh	5	Alisha	2
Kirsten	5	Brian	2

These data are not outliers, but now there are several students who watched only a few movies in one month. What is the mean of the data now? Compare the means you found in parts A, B, and C with this new mean. What do you notice? Why?

Assignment Choices

ACE questions 8 and 9 and unassigned choices from earlier problems

Assessment

It is appropriate to use Check-Up 2 and the Quiz after this problem.

Answers to Problem 5.5

A. The total number of students is 10. The total number of movies watched is 90. The mean number of movies watched is 9.

B. See page 67n.

C. See page 67n.

D. See page 67n.

■ **Problem 5.5 Follow-Up**

1. What happens to the mean when you add one or more values that are much larger than the values in the original data set? Why does this happen?

2. What happens to the mean when you add a number of values that are clumped with the smaller values in the original data set? Why does this happen?

Answers to Problem 5.5 Follow-Up

1. Adding unusual values to the data set can greatly affect the mean. Adding very large numbers pulls the mean up (increases it).

2. Adding several small numbers pulls the mean down (decreases it).

Answers

Applications

1a. 3; Order the data from smallest to largest. The median is the value that separates the data in half.

1b. Yes, 6 of the families have 3 children. This is possible because the median is located using the data values. The only time the median will not be one of the data values is when it is determined by finding the mean of two middle values that are not identical.

2a. 4; You can add the data values together and divide by the number of data values to get the mean. Or, you can find the mean by making cube towers for each of the households and then evening out the towers so there are 16 households, each with 4 members. The mean tells you the value that each data item would have if all the data had the same value.

2b. There are no squares over the number 4 on the line plot, which means there are no households in the data set with four children. This is possible because there are households with more than four children and households with less than four children to balance each other.

3a. 48

3b. See below right.

As you work on these ACE questions, use your calculator whenever you need it.

Applications

In 1 and 2, use the line plot below, which shows information about the number of children found in 16 households.

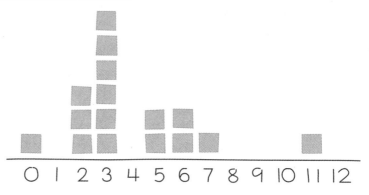

1. **a.** What is the median number of children in the 16 households? Explain how you found the median and what it tells you.

  b. Do any of the 16 households have the median number of children? Explain why this is possible.

2. **a.** What is the mean number of children in the 16 households? Explain how you found the mean and what it tells you.

  b. Do any of the 16 households have the mean number of children? Explain why this is possible.

3. The mean number of people in eight households is 6.

  a. What is the total number of people in the eight households?

  b. Make a line plot showing one possible arrangement for the numbers of people in the eight households.

3b. Possible answer:

```
                    X
     X      X  X  X  X  X      X
    _____
     1  2  3  4  5  6  7  8  9  10  11
```

c. Make a line plot showing a different possible arrangement for the numbers of people in the eight households.

d. Are the medians the same for the two arrangements you made?

4. A group of nine students has a mean of $3\frac{1}{2}$ people in their households. Make a line plot showing an example of this data set.

5. A group of nine students has a mean of 5 people in their households, and the largest household in the group has 10 people. Make a line plot of a data set fitting this description.

Connections

6. Jon has a pet rabbit that is 5 years old. He wonders if his rabbit is old compared to other rabbits. At the pet store, he finds out that the mean age for a rabbit is 7 years.

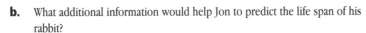

a. What does the mean tell Jon about the life span for a rabbit?

b. What additional information would help Jon to predict the life span of his rabbit?

7. A store carries nine different brands of granola bars. What are possible prices for each of the nine brands of granola bars if the mean price is $1.33? Explain how you determined the values for each of the nine brands. You may use pictures to help you.

3c. Possible answer:

```
        X
        X           X
    X   X   X   X   X
    _____
    3   4   5   6   7   8   9
```

5. Possible answer:

```
        X       X
    X   X   X   X   X   X           X
    _____
    1   2   3   4   5   6   7   8   9  10  11
```

3c. See below left.

3d. Answers will vary.

4. This situation is impossible. For nine households to have a mean of $3\frac{1}{2}$ people, there would have to be a total of $9 \times 3\frac{1}{2}$, or $31\frac{1}{2}$ people.

5. See below left.

Connections

6a. The mean tells Jon that if all the rabbits in the data set that was used to find the mean lived to be the same age, that age would be 7 years. What actually happens is that some of the rabbits don't live to 7 years and some of the rabbits live beyond 7 years.

6b. Knowing the range would give Jon more information about the possible life span of his rabbit.

7. If the mean price of a box of granola bars is $1.33 and there are nine different brands of granola, the total price for one box of each brand is $11.97. You could have the nine brands all priced at $1.33, or have just a few priced at $1.33, or have no brands priced at $1.33. Here is one possibility: $1.35, $1.39, $1.49, $1.17, $1.29, $1.35, $1.25, $1.29, $1.39

8a. Mayor Phibbs determined the mean income. The total of the incomes is $32,000; dividing by the number of incomes, 16, gives $2,000 per week. Louisa Louis found the median income. There are a total of 16 values, so the median is between the eight and ninth value. The eighth value is 0 and the ninth value is 200, so the median is 100. Radical Ronnie Radford looked at the mode income, which is 0. Each of their computations is correct.

8b. no; No one earns $2,000 per week.

8c. no; No one earns $100 per week.

8d. yes; Eight people earn $0 per week.

8e. $200 is a good answer. Possible explanation: The people who have $0 incomes are probably children, so the people who earn $200 and the person who earns $30,600 are the residents who are employed. The "typical" income would be either the median or the mode of the incomes of $200 or more (both the median and the mode are $200); however, the mean would be greatly affected by the one large income.

8f. The mode would be $200. The median would be $200. The mean would be $1,640.

9a. See right.

9b. 9 movies

---

8. Three candidates are running for the mayor of Slugville. Each has determined the typical income for the people in Slugville and is using this information to help in the campaign.

Mayor Phibbs is running for re-election. He says, "Slugville is doing great! The average income for each person is $2000 per week!"

Challenging candidate Louisa Louis says, "Slugville is nice, but it needs my help! The average income is only $100 per week."

Radical Ronnie Radford says, "No way! We must burn down the town—it's awful. The average income is $0 per week."

None of the candidates is lying. Slugville has only 16 residents, and their weekly incomes are $0, $0, $0, $0, $0, $0, $0, $0, $200, $200, $200, $200, $200, $200, $200, and $30,600.

a. Explain how each of the candidates determined what the "average" income was for the town. Check the computations to see whether you agree with the three candidates.

b. Does any person in Slugville have the mean income? Explain.

c. Does any person in Slugville have the median income? Explain.

d. Does any person in Slugville have the mode income? Explain.

e. What do you consider to be the typical income for a resident of Slugville? Explain.

f. If four more people moved to Slugville, each with a weekly income of $200, how would the mean, median, and mode change?

9. A recent survey asked 25 middle-school students how many movies they watch in one month. The data are shown on the next page. Notice that the data are quite spread out; the range is from 1 to 30 movies.

a. Make a stem-and-leaf plot to show these data. Describe the shape of the data.

b. Find the mean number of movies watched by the students for the month.

9a. Possible answer: The data are bunched in the intervals of 0–9 and 10–19, with one value at 20, one at 25, and one at 30.

```
0 | 1 1 1 2 2 2 3 3 4 4 4 5 8 8 9
1 | 0 1 2 3 5 5 7
2 | 0 5
3 | 0
```

c. Describe how you found the mean number of movies.

d. What do the mean and the range tell you about the typical number of movies watched for this group of students?

e. Find the median number of movies watched. Are the mean and the median the same? Why do you think this is so?

Movies Watched by Students

Student	Movies per month
Wes	2
Sanford	15
Carla	13
Su Chin	1
Michael	9
Mara	30
Alan	20
Brent	1
Tanisha	25
Susan	4
Darlene	3
Eddie	2
Lonnie	3
Gerald	10
Kristina	15
Paul	12
Henry	5
Julian	2
Greta	4
TJ	1
Rebecca	4
Ramish	11
Art	8
Raymond	8
Angelica	17

9c. Add to find the total number of the movies watched (225), then divide the total by the number of students (25).

9d. The range is 1 to 30 movies and the mean is 9 movies. Since the mean is more toward the low end of the range, more students fall in the low end of the range.

9e. The median number of movies watched is 8. The mean is larger than the median because the large values pull the mean up, but have less influence on the median.

Extensions

10. Answers will vary. Pay attention to the students' reasoning. Generally, data reported in newspapers use the mean.

11. There are 365 days in a year. This means the average third grader watches $3\frac{1}{5}$ hours of television a day.

12a. Ms. R's Class
Mean: $34\frac{9}{14}$, ($970 \div 28$)
Median: $26\frac{1}{2}$

Mr. K's Class
Mean: $54\frac{2}{5}$ ($1632 \div 30$)
Median: 34

The mean is larger than the median for each class, because there are some larger values in each set of data.

12b. They probably would want to use the mean, because their mean is about 20 jumps larger. Their median is about 8 jumps larger.

12c. The median increases by 1 to 33 jumps. The median is the middle data value, so it is not changed much by removing the largest data value.

12d. The mean is now ($1332 \div 29$) or $45\frac{27}{29}$ or 45.931. The mean decreases by a little more than 8 jumps. It decreases because the largest value was removed, which had a large influence on the mean.

12e. Ms. Rich's class's mean and median are still less than each of the same statistics for Mr. Kocik's class, so Ms. Rich's class cannot make a valid claim that they are better.

Extensions

In 10 and 11, consider this newspaper headline:

10. Which average—median or mean—do you think is being used in this headline? Explain why you think this.

11. About how many hours per day does the average third grader watch television if he or she watches 1170 hours in a year?

12. Review the jump-rope data for Problem 3.2 on page 35.

   a. Compute the median and the mean for each class's data. How do the median and the mean for each compare?

   b. Which statistic—the median or the mean—would Mr. Kocik's class want to use to compare their performance with Ms. Rich's class? Why?

   c. What happens to the median of Mr. Kocik's class's data if you leave out the data for the student who jumped rope 300 times? Why does this happen?

   d. What happens to the mean of Mr. Kocik's class's data if you leave out the data for the student who jumped rope 300 times? Why does this happen?

   e. Can Ms. Rich's class claim to be better if the data of 300 jumps is left out of Mr. Kocik's class's data? Explain why or why not.

Mathematical Reflections

In this investigation, you explored a type of average called the mean. You used cubes to help you see what it means to "even out" data to locate the mean, and you created different data sets with the same mean. Then you developed a way to find the mean without using cubes. Finally, you looked at what happens to the mean when the data include very high or very low values. These questions will help you summarize what you have learned:

1 Describe a method that requires using only numbers for finding the mean. Explain why this method works.

2 You have used three measures of center: the mode, the median, and the mean.

    a. Why do you suppose these are called "measures of center"? What does each tell you about a set of data?

    b. Why might people prefer to use the median instead of the mean?

3 You have used one measure of spread: the range.

    a. Why do you suppose the range is called a "measure of spread"?

    b. Why might people prefer to describe a data set using both a measure of center and a measure of spread rather than just one or the other?

4 Once you collect data to answer questions, you will want to decide what measures of center and spread can be used to describe your data.

    a. One student said she could use only the mode to describe categorical data, but that she could use the mode, median, and mean to describe numerical data. Is she right? Explain why or why not.

    b. Can you determine a measure of spread for categorical data? Explain.

Think about your answers to these questions, discuss your ideas with other students and your teacher, and then write a summary of your findings in your journal.

You will soon be developing your own survey to gather information about middle-school students. What measures of center and spread can you use to describe the data you might collect for each question in your survey?

1. Add all of the values. Divide the sum by the number of values. This gives the mean. This works because the sum of the values is the amount to be shared or "evened out." The number of values is the number of parts into which the total must be divided. Division gives the number in each part.

2a. They are measures of center because they generally fall where most of the data cluster. The mode is the data value that occurs most frequently. The median is the middle value, which separates the ordered set of data in half. The mean is the "balance point," or the value that each item would have if all the data had the same value.

2b. The median is not affected by extremes—by large or small values—in the data.

3a. It tells you the smallest and largest values in the data, or how spread out they are.

3b. These two measures give the upper and lower boundaries of the data set, and the point where most of the data cluster.

4a. See left.

4b. See left.

4a. She is correct. The mean and median are measures that can only be used with numerical data. The mode, a count of the most frequently occurring data value within a data set, can be used with both numerical and categorical data.

4b. no; A measure of spread depends on being able to identify the smallest and largest values. You cannot order categorical data in a logical way.

5.1 • Evening Things Out

Launch

Defining how to count or measure is a critical part of the process of statistical investigation. Initial decisions in this area impact the outcome of many statistical studies. The United States census is a survey that seeks to count the number of people living in the United States and to describe key characteristics about these people, such as household size. Students use the definition for *household* provided by the United States census throughout Investigation 5.

> Does anyone know what the purpose of the United States census is?

Historically, the purpose of the census is to count the number of people living in the United States. This is done to compute the number of representatives each state will have in the United States House of Representatives.

> The Census Bureau has developed a definition of who to count. The census focuses on counting the people who live in households rather than asking questions like "How many people are in your family?"

> The word *household* refers to all the people who live in a "housing unit," which may be a house, an apartment, some other group of rooms, or a single room, like a room in a boardinghouse.

> Why do you think the census asks "How many people are in your household?" and not "How many people are in your family?"

By defining the word *household* as it does, the census seeks to eliminate confusion about who to count. Only the people who live in a household and use that location as their permanent address at the time the census is taken are counted as members of that household. When you try to define the word *family*, you run into a lot of questions about who to count: "What if I have an older sister who doesn't live with us now?" "Can I count my grandmother?" "How do I count my stepsister and stepbrother?" The census must follow a single rule about who to count. The definition for household helps census takers to report population data accurately.

Explore

We suggest you do Problem 5.1 through class instruction. The material in the student edition may be used as reference for students during or after instruction.

Work with students to consider several different responses to the question, What are some ways to determine the average number of people in these eight households? If students do not suggest it themselves, have them consider using the mode and median as two possibilities. If students suggest using the mean by saying something such as, "Well, you can add up all the numbers and divide by 8," say you will come back to that idea shortly.

Once students have explored finding the mode (6 people) and the median ($3\frac{1}{2}$ people) as ways to describe the average number of people in the households, introduce the idea of another way to think about average.

We suggest you have students construct the physical model—in small groups—by building a tower of cubes for each student's household. Each tower should be a single color and different from the colors of the other towers (this way, you can refer to "Ollie the orange tower," "Ruth the red tower," or "Brenda the blue tower"). The towers should look something like this:

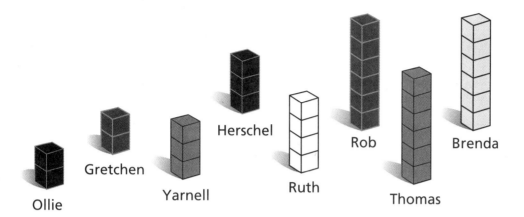

Have the groups arrange their cube towers in order from smallest to largest.

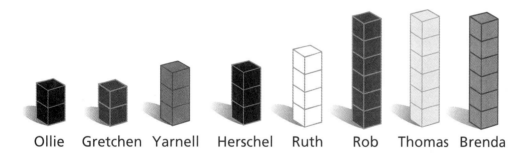

These students decided to find the average by "evening out" the number of cubes in each tower. Try this on your own and see what you come up with.

After students have had a few minutes to work with their towers, have a class discussion about what they found out. The towers are "evened out" by moving cubes from taller towers to shorter towers. Since each tower was originally a single color, students will be able to see from which towers the "moved data" came.

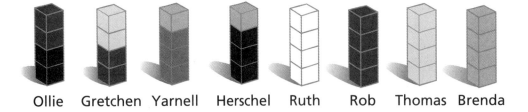

Ollie   Gretchen Yarnell   Herschel   Ruth   Rob   Thomas Brenda

We can say that each household has an average of four people. Some of the households (Rob's, Thomas's, and Brenda's) actually have more than four people; their extra cubes have been moved to households with fewer than four people (Ollie's, Gretchen's, Yarnell's, and Herschel's). One household (Ruth's) already had four people, so none of her cubes were moved.

We call this "evened out" number the *mean*. How does the mean compare to the mode and the median for this set of data?

For the Teacher

This situation was designed so that the mode, median, and mean are all different. Note, however, that the median and mean are quite similar.

After students understand the tower representation and have discussed the idea of evening out, introduce the line-plot representation.

> Using stick-on notes and a number line, let's now make a line plot to show these data.

Rebuild the original cube towers. Begin to build the line plot by indicating that the first person, Ollie, has a two-person household, so one stick-on note is placed above the 2 on the line plot. Work back and forth between the cube representation and the line-plot representation to complete the model.

Numbers of People in Households

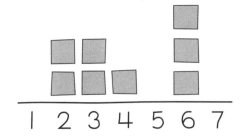

1 2 3 4 5 6 7

Students may not easily see that the cube towers and the line plot display the same information. Help them to explore the models by asking such questions as these:

Looking at the towers, how many households are shown? (*8*)

Looking at the line plot, how many households are shown? (*8*)

Looking at the tower, how many people are in the eight households altogether? (*32*)

How did you figure this out? (*by counting the number of cubes*)

How should the number of people represented by the line plot compare to the number of people represented by the towers? (*It should be the same, as it was built using the data from the towers.*)

How can you verify your thinking?

Help students to understand how to compute the total number of people from the line plot: the two stick-on notes above the 2 mean two households of two people, or a total of four people; the two stick-on notes above the 3 mean two households of three people, or a total of six people; and so on.

The concept that a stick-on note above a numeral on the line plot represents that number of people in one household may not be easy for students to grasp. With the cube model, each tower represents the number of people in a household, so the cubes make it easier for students to visualize how to count the total number of people.

Summarize

Return to the line plot with the stick-on notes to relate the "evening out" model back to the line-plot representation. First, discuss the mean as the balance point in the distribution.

When we moved the cubes to even out the towers, the towers for households with fewer than four people were made taller, and the towers for households with more than four people were made shorter.

We moved two cubes from each of the three towers with six cubes. So we decreased our towers by a total of six cubes. (*Write 2 + 2 + 2 = 6 on the board.*)

We added two cubes to each of the two towers with two cubes, and we added one cube to each of the two towers with three cubes. So we increased the towers by a total of six cubes. (*Write 2 + 2 + 1 + 1 on the board.*)

Students will be able to see these changes by observing where the colors ended up after the evening out was completed.

The amount of increase was the same as the amount of decrease. Now let's see how all of this relates to our line plot.

Show the top portion of Transparency 5.1A.

Here is our line plot. The arrow points to the mean. You can think of the line plot as balancing on the arrow.

Numbers of People in Households

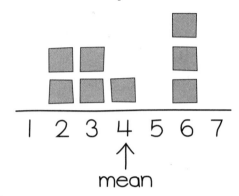

To even out the values, we took 2 away from each data value of 6 to get to the mean of 4. This makes our distribution out of balance.

Reveal the second graph on Transparency 5.1A.

Numbers of People in Households

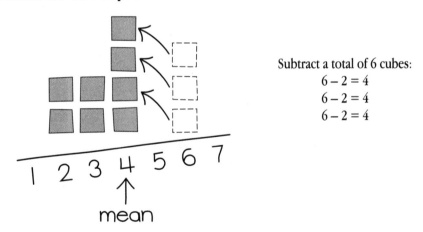

Subtract a total of 6 cubes:
$$6 - 2 = 4$$
$$6 - 2 = 4$$
$$6 - 2 = 4$$

To balance the distribution, we need to increase the data values of 2 and 3 to bring them up to the mean of 4.

Show the third graph on Transparency 5.1A.

By comparing the equations next to each line plot, you can see that the total number of cubes we subtract is the same as the total number of cubes we add.

Numbers of People in Households

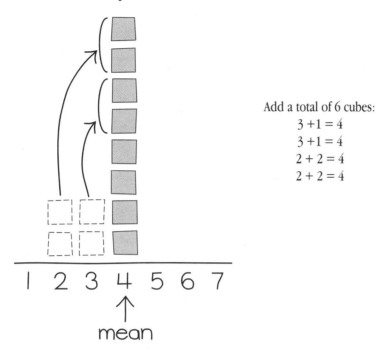

Add a total of 6 cubes:
$$3 + 1 = 4$$
$$3 + 1 = 4$$
$$2 + 2 = 4$$
$$2 + 2 = 4$$

Our cube towers let us even out the numbers in each household to find the mean. Now we see how we can "even out" the households on our line plot to show the mean as a balance point.

Have students try to write a definition of the term *mean* in their journals. Their definitions may be vague now; you will return to this task in Problem 5.4.

5.2 • Finding the Mean

Launch

This problem is a continuation of Problem 5.1. Provide pairs of students with interlocking cubes and stick-on notes.

Discuss the problem briefly so all the students understand what they are to do.

Explore

As you move around the room, help pairs to understand that they need to build eight towers, one tower for each of the eight students, with the numbers of cubes that represent the members of each household. Look for line plots that match the data.

Summarize

Discuss each of the questions in Problem 5.2, encouraging students to share their strategies for solving the problem. Most students will use the strategy of evening out the cubes to find the mean, but they may use other strategies as well. However, they need to justify their strategies by explaining why what they did works.

5.3 • Data with the Same Mean

Launch

Refer students to the line plots on page 58 of the student edition, or display Transparency 5.3.

> Look at the line plots for the data in Problems 5.1 and 5.2. How are these distributions alike? How are they different?

> How many households are there in each situation? (*8*)

> How many total people are there in each situation? (*32*)

> How are these facts related to the mean of 4 in each case?

Students will notice that the distributions look different and that the data values are different. However, they both show 8 households with a total of 32 people and a mean of 4 people. Some students may observe that dividing 32 people among 8 households is 4 people per household.

Propose the idea of having additional data sets with the same mean.

> Do you think it is possible to have other sets of data about eight households that are different from the ones we explored and still have a mean of 4 people?

Hopefully, students will think this is possible. Some may not agree; some may not know. Challenge students to work with you to make a new distribution with eight households and a mean of 4 people.

> Our data set must have a total of eight households. Suppose that the first household had three people. I can show these three people by making a three-cube tower.

Make a tower of three cubes of the same color.

> How many more towers do I need to make to show the remaining households? (*7*)

Now, suppose another household has eight people. I'll make a tower of eight cubes to represent that household.

Using a different color, make a tower of eight cubes.

How many people are represented with just the two towers we've made? (*11*)

What do you think I should do next?

Some students may see that the goal is to have a total of 32 people in the 8 households and that, since we have accounted for 11 people, the remaining 6 households must have 21 people altogether. As you construct each new tower, students will need to count how many more of the 32 people have been accounted for. Be patient with this discussion. Once you have a set of 8 towers, you may want to make a line plot and then check for a mean of 4 by evening out the towers. If the mean is not 4, discuss with students what you need to do to the data set so that the mean will be 4.

Explore

Have pairs work on Problem 5.3. For each data set, they create a line-plot distribution with stick-on notes. Post the line plot for one distribution from each team for the whole class to see. Do a quick check as you scan their line plots: the total number of people represented must be 32.

Summarize

Ask students to look at the displays for part A.

Are any of the distributions the same? How many different distributions are posted?

Point to specific examples, and ask questions about them.

How many households are represented? (*8*)

How many people are there in all the households? (*32; make sure students compute this number*)

What is the mean? (*4*)

Discuss part B as thoroughly as you did part A. Post distributions, and ask the same questions about the number of households, the total number of people, and the mean.

Have pairs or small groups work on Problem 5.3 Follow-Up.

You will want to help students synthesize what they know about making distributions with different means. Ask questions that help them focus their observations on identifying common strategies.

What are some quick ways to come up with different sets of data with eight households and a mean of 4 people?

What are some quick ways to come up with different sets of data with nine households and a mean of 4 people?

What are some quick ways to come up with different sets of data with nine households and a mean of 3 people?

What are some quick ways to make different sets of data with eight households and a mean of $3\frac{1}{2}$ people?

Ideally, students will see that if they know the number of households and the mean number of people in the households, they can determine the number of people in all of the households. Using this total, they can work backward to create a data set showing the total number of people distributed among the households.

How can you use what you know about the number of people in all the households and the number of households to find the mean number of people?

This question leads into Problem 5.4. Students have been working with facts about numbers of households and the mean number of people in those households. This question asks them to think about the problem in a slightly different way.

Students have used the number of households and the mean to determine the total number of people:

Known: number of households, mean number of people
Unknown: total number of people

Now, we want them to use the number of households and the total number of people to find the mean:

Known: number of households, total number of people
Unknown: mean number of people

Pose an example:

Suppose there are six households with a total of 36 people. What is the mean number of people in each household?

Have students use interlocking cubes to build six towers representing the possible numbers of people in each household. Work with them as they even out the towers to find the mean. Can they answer your original question now? Then, ask them to find the mean in a different way.

How could you determine the mean without using cubes?

Some students may visualize distributing 36 cubes evenly among the six households. This is like finding $36 \div 6$. Other students may use the example of the evened-out cubes, noting that the total numbers of people and households stay the same; we just have to share the cubes among all the households. Students may suggest other strategies.

5.4 • Using Your Class's Data

Launch

The Summarize phase of Problem 5.3 is an excellent lead-in to Problem 5.4. Continue the line of questioning begun at the end of Problem 5.3, using the following examples:

> There are 12 students with a total of 60 people in their households. How would we go about finding the mean number of people in each household? What is the mean number of people in each household? (5) Why does your strategy work?

> There are 50 students with a total of 225 people in their households. What is the mean number of people in each household? ($4\frac{1}{2}$) Explain your strategy and why it works.

Explore

Present Problem 5.4 to your students. Have them work in pairs or small groups to collect the class's data, make a display to show the information, and calculate the mean number of people in their households.

Students may use different strategies for finding the mean. The goal is for them to be able to compute the mean using the algorithm and to be able to explain how their computations mirror work they could do using cubes.

Problem 5.4 Follow-Up gives a model for writing a summary of their work. Have each team prepare a similar written summary of Problem 5.4 in their journals.

Summarize

Have students return to the definition of *mean* they wrote after Problem 5.1 and refine it. Have groups tell what strategy they used to find the class mean. Ask why they think they are correct. Focus on meaning.

> What does the class mean tell us? How is a mean different from a median?

You want the students to see that the two measures are both kinds of "middles." The median is the physical middle of the values and the mean is the balance point value where the distribution would be evened out if all the values were equally shared.

5.5 • Watching Movies

Launch

The data used in this problem were collected from a group of middle-school students who answered this question: How many movies did you watch last month?

There were some misunderstandings about what "movie" meant in this context. It seems the students responding to the survey thought of "movie" as being any kind of video or television movie they watched during the previous month. Discuss the question and possible ways students might interpret what is being asked. How would your students answer this question? What kinds of things would they want to clarify?

Your students might want to think about the mean number of movies a day that some students in this group watched. For example, the student who watched 42 movies in one month averaged $1\frac{2}{5}$ movies a day!

Explore

Have pairs work together to complete the problem.

Summarize

Use Problem 5.5 to focus the summary.

> What was the mean in part A of 5.5? In part B, a student whose data is very much larger than the rest of the data is added. What happened to the mean?
>
> In part C, we added Tamara whose data is very large! What happened to the mean?
>
> In general, what effect do you think large outliers have on the mean of a data set? Why?
>
> Let's test our ideas on our data.
>
> What data values could you add to cause the mean of the 9 movies to increase? (*values that are greater than 9*)
>
> What data values could you add to cause the mean of 9 movies to decrease? (*values that are less than 9*)
>
> What data values could you add to cause the mean of 9 movies to remain the same? (*values that are less than 9 paired with values that are greater than 9, along with any number of values equal to 9*)

For the Teacher

It is appropriate to use the Quiz after Problem 5.5. The day before administering the quiz, ask each student to write his or her bedtime on a sheet of paper and pass it to you. Display this data (without names) during the quiz.

Additional Answers

Answers to Problem 5.2

A.

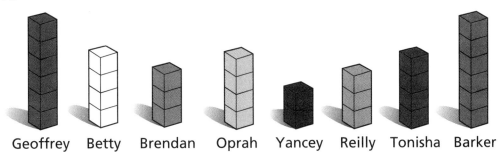

Geoffrey Betty Brendan Oprah Yancey Reilly Tonisha Barker

B.

Numbers of People in Households

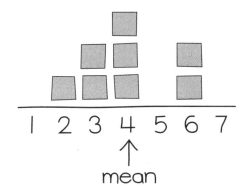

Answers to Problem 5.3

A. Possible line plots:

```
                    X
              X     X
        X  X  X  X        X
       ─────────────────────
        1  2  3  4  5  6  7  8
```

```
        X           X
        X           X
        X           X
        X           X
       ─────────────────────
        1  2  3  4  5  6  7  8
```

```
           X
           X
        X  X  X  X           X        X
       ──────────────────────────────────
        1  2  3  4  5  6  7  8  9  10  11
```

B. Possible line plots:

```
        X           X       X
    X   X   X   X   X       X
    _____
    0   1   2   3   4   5   6   7   8


        X               X
        X               X
        X   X   X   X   X
    _____
        1   2   3   4   5   6   7


            X   X   X
            X   X   X
            X   X   X
        _____
            2   3   4   5   6
```

Answers to Problem 5.3 Follow-Up

1. Possible line plots:

```
                X
                X
            X   X
        X   X   X   X           X
    _____
        0   1   2   3   4   5   6   7


                X       X
                X       X
                X       X
                X   X   X
        _____
            0   1   2   3   4   5


                X
            X   X
            X   X
        X   X   X                   X
    _____
        0   1   2   3   4   5   6   7   8   9
```

2. Possible line plots:

```
            X           X
            X           X
            X   X   X   X
        _____
            1   2   3   4   5   6


                X
            X   X   X
            X   X   X           X
        _____
            1   2   3   4   5   6   7
```

```
        X           X
    X   X   X   X   X   X
  ──────────────────────────
  0   1   2   3   4   5   6   7
```

Answers to Problem 5.5

B.   Add 42 to the stem plot:

```
0 │ 3 3 5 6 6 7
1 │ 1 5 6 8
2 │
3 │
4 │ 2
```

The new mean is 12 movies. It is larger than the mean for the data in part A.

C.   Add 96 to the stem plot:

```
0 │ 3 3 5 6 6 7
1 │ 1 5 6 8
2 │
3 │
4 │ 2
5 │
6 │
7 │
8 │
9 │ 6
```

The new mean is 19 movies. The value of 96 made the mean larger than either of the two earlier means.

D.   Add the eight new values to the stem plot:

```
0 │ 2 2 3 3 4 4 5 5 5 5 5 6 6 7
1 │ 1 5 6 8
2 │
3 │
4 │ 2
5 │
6 │
7 │
8 │
9 │ 6
```

The new mean is 13 movies. It has decreased from the mean of 19 movies in part C. However, the two higher values of 42 and 96 movies still influence where the mean occurs.

Assigning the Unit Project

The Is Anyone Typical? Project was introduced at the beginning of the unit and is formally assigned here. Students are asked to use what they have learned in *Data About Us* to conduct a statistical investigation to determine some typical characteristics of middle-school students.

Some schools may require administrative approval of surveys; check prior to data collection. It may also be a problem for several classes to conduct surveys independently. You may want to coordinate the data collection among all of the classes so classes are not disturbed several times.

The project can be assigned in a variety of ways. If you have several days available, you can have each group write and conduct a survey consisting of from five to ten questions. Each group then collects, analyzes, and interprets the data and prepares a report of their findings. If your time is limited, you may choose to work as a class to develop and conduct the survey. You can then have each group analyze and interpret the data for one question.

A detailed discussion of the project, samples of student projects, and a suggested scoring rubric are given in the Assessment Resources section.

Is Anyone Typical?

You can use what you have learned in *Data About Us* to conduct a statistical investigation to answer the question, "What are some characteristics of a typical middle-school student?" When you have completed your investigation, make a poster, write a report, or find some other way to communicate your results.

Your statistical investigation should consist of four parts:

Posing questions
You will want to gather both numerical data and categorical data. Your data may include physical characteristics, family characteristics, miscellaneous behavior (such as hobbies), and preferences or opinions. Once you have decided what you want to know, you need to write appropriate questions. Make sure that your questions are clear so that everyone who takes your survey will interpret them in the same way.

Collecting the data
You may want to collect data from just your class or from a larger group of students. You also need to decide how to distribute and collect the survey.

Analyzing the data
Once you have collected your data, you need to organize, display, and analyze them. Be sure to think about what kinds of displays and which measures of center are most appropriate for each set of data values you collect.

Interpreting the results
Use the results of your analysis to describe some characteristics of the typical middle-school student. Is there a student that fits all the "typical" characteristics you found? If not, explain why.

Tips for the Linguistically Diverse Classroom

Chart Summary The Chart Summary technique is described in detail in *Getting to Know CMP*. This technique involves presenting information by condensing it in a pictorial chart with minimal words. Example: Use a four-part pictorial chart as you present the steps of a statistical investigation. For each part, draw rebuses of stick figures carrying out the activities.

Assessment Resources

Check-Up 1

1. Consider each distribution below. For each distribution, where possible, tell how many people are represented by the data, and identify the mode, median, and range.

   a.

Lengths of Last Names

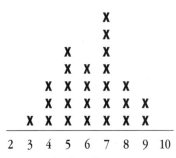

```
                    X
                    X
              X     X
              X  X  X
           X  X  X  X  X
           X  X  X  X  X  X
        X  X  X  X  X  X  X
        ─────────────────────────
        2  3  4  5  6  7  8  9  10
              Number of letters
```

   b.

Birth Months

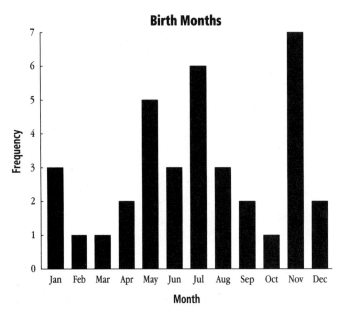

2. On the back of this page, make a line plot showing the lengths of 11 names so that the median length is 12 letters and the range is from 6 letters to 16 letters.

3. The media specialist in your school is planning a book fair. She is preparing a survey to ask students a few questions to help make the book fair a success.

   a. Write one question that will give the media specialist *numerical* data.

   b. Write one question that will give the media specialist *categorical* data.

Quiz

Questions 1–3 explore the question of what time a typical student goes to bed on a school night. Your teacher will share your class's data on bedtimes with you. Use this data for the following questions.

1. Organize and display your class's data on bedtimes using an appropriate graph.

2. Answer the following questions about your class's data.
   a. What is the range of the data?

   b. What is the mode of the data? How many students had a bedtime the same as the mode?

   c. What is the median of the data? How many students had a bedtime the same as the median?

   d. Are there any outliers in the data? If so, what might cause this?

3. Based on the class's data, what would you say is the typical time a student in your class goes to bed on a school night? Explain your reasoning.

Quiz

4. A group of students were curious about the changes in people's height over time. They gathered data about height from two different groups of students in their district: students in grade 5 and students in grade 8. The data they collected is shown in the table.

   a. Make stem-and-leaf plots that show these data.

   b. What is the typical height of the grade 5 students? Justify your answer.

   c. What is the typical height of the grade 8 students? Justify your answer.

   d. How does the height data from the grade 5 class compare with the height data from the grade 8 class?

   e. There were three grade 8 students absent the day the data were collected. Their heights are 177 cm, 187 cm, and 163 cm. What happens to the mean, mode, and median when these new values are added to the data set?

Height (centimeters)	
Grade 5	Grade 8
138	147
138	156
138	159
139	160
141	160
142	161
144	162
146	162
147	162
147	162
147	163
150	164
150	165
151	165
151	168
151	168
151	168
152	168
152	169
152	171
152	172
153	174
153	176
155	
155	
156	
156	
157	
158	
171	

Quiz

5. Two students went to a frog-jumping contest. They wondered whether there might be a relationship between a frog's weight and its jumping ability. The biology teacher had frogs in her lab, and the two students decided to investigate their question: "What is the relationship between a frog's weight and how far it jumps?" They collected the data on the right from the 26 frogs in the science lab.

a. On another sheet of paper, make a coordinate graph that shows each frog's weight and the length of its jump. Put weight measurements on the *x*-axis (the horizontal axis) and length measurements on the *y*-axis (the vertical axis).

b. Describe any patterns in the graph.

c. What can you say about the relationship between the weight of a frog in this group and the length of its jump?

d. What is the median weight of the frogs?

e. What is the median jump of the frogs?

f. How many frogs have a weight above the median weight and a jump above the median jump?

g. How many frogs have a weight above the median weight and a jump below the median jump?

Frog Data

Name of frog	Weight (grams)	Length of jump (centimeters)
Jumper	70	43
Webster	100	64
Leapy Leo	80	32
Kroaker	120	46
Fruity	100	37
Frogzilla	120	46
Jalapeño	100	28
Big Bertha	70	52
Tommy	90	34.5
Thunder	140	49.5
Kirby	120	35.5
Sliminator	120	30
Horton	90	26
Kekokekory	100	29
Pippin	130	52
Speedy	110	26.5
Lightning	80	32
Coco	135	32
Fast Freddie	120	27.5
Jumpfaster	100	26
Terminator	130	26
Froggy	100	54
Kickin'	130	49
Sir Kermit	130	30
Bullet	130	34

Check-Up 2

1. A group of 9 students has these numbers of children in their families: 3, 2, 4, 2, 1, 5, 1, 2, and 7.
   a. Find the median number of children in the 9 families.

   b. Find the mean number of children in the 9 families.

2. The stem plot below shows test scores for Ms. McIntyre's class on a state mathematics test. Students could score from 0 to 100 points.

   Class Test Scores

   ```
   0 | 5
   1 |
   2 | 4
   3 | 4 9
   4 | 3 7 8
   5 | 7 9
   6 | 1 6 8
   7 | 3 5 6 8 8
   8 | 1 2 2 2 5
   9 | 0 3 9
   ```

   a. Are these data numerical or categorical?

   b. What is the range of the data?

   c. What is the median of the data? How many students had a score the same as the median?

Check-Up 2

3. Fifteen students read the book *Gulliver's Travels*. In the book, the Lilliputians said they could make clothes for Gulliver by taking one measurement, the length around his thumb. The Lilliputians claimed that

   • the distance around Gulliver's wrist would be twice the distance around his thumb.

   • the distance around Gulliver's neck would be twice the distance around his wrist.

   • the distance around Gulliver's waist would be twice the distance around his neck.

   The students wondered whether this doubling relationship would be true for them too. They measured the distance around their thumbs and their wrists in centimeters, then graphed the pairs of numbers on a coordinate graph. They drew a line connecting the points that represented wrist measurements that were twice thumb measurements.

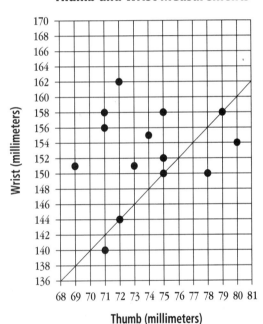

Thumb and Wrist Measurements

   a. How many students' measurements fit the Lilliputian rule that twice the distance around the thumb equals the distance around the wrist?

   b. How many students' wrist measurements are less than twice their thumb measurements?

Check-Up 2

c. The point for Jeri's thumb and wrist measurements is above the line. If the cuffs of a shirt are twice the measurement around Jeri's thumb, how will the cuffs of the shirt fit her?

d. The point for Rubin's thumb and wrist measurements is below the line. If the cuffs of a shirt are twice the measurement around Rubin's thumb, how will the cuffs of the shirt fit him?

4. Suppose you wanted to describe the typical student in your grade. You decide to design a survey to help collect information about students in your grade.

   a. Write one question for your survey that will give you *numerical* data. Explain how this information would help you to describe the typical student.

   b. Write one question for your survey that will give you *categorical* data. Explain how this information would help you to describe the typical student.

Assign these questions as additional homework, or use them as review, quiz, or test questions.

1. For the distribution below, tell how many people are represented and identify the mode, median, and range.

Lengths of First Names

```
X
X  X
X  X  X
X  X  X  X
────────────────────
3  4  5  6  7  8
```
Number of letters

2. A class investigated the question of how many paces it takes to travel from their class to the gym. They measured the distance by counting the number of paces each student walked. Every step made on the right foot counted as one pace. Here are their results:

Paces to the Gym

```
                  X
                  X
                  X  X
                  X  X
        X         X  X  X
X                 X  X  X  X
X  X  X  X  X  X  X  X  X
────────────────────────────────
16 17 18 19 20 21 22 23 24 25 26
```

   a. What is the median number of paces the students took to travel the distance?

   b. Make a bar graph that displays this information. Explain how the bar graph is similar to and different from a line plot.

   c. Who has the shorter pace: the student who traveled the distance in 17 paces or the student who traveled the distance in 25 paces? Explain your reasoning.

3. Make a line plot showing the ages in years of 12 students so that the median age is 12.5 years and the difference between the highest age and the lowest age is 9 years.

4. The mean number of children in six families is 5 children.

   a. What is the total number of children in the six families?

   b. Other than six families of 5 children, create a set of families that fits this information.

   c. Would another classmate's set of families for question b have to be the same as yours? Explain.

© Dale Seymour Publications®

5. In the story *The Phantom Tollbooth,* Milo is told that the average number of children in a family is 2.58. You know that a .58 boy or girl cannot exist. How could the calculations for the mean produce this number?

6. Most people will walk about 158,125 kilometers in their lifetime, or around the world 4.5 times.

   a. How do you suppose this statistic was determined?

   b. What might you do if you were asked to investigate the question, How far do most people walk in their lifetime?

7. A class investigated how many pets each student in the class had. A number of students in their class had no pets at all. Here's how their data looked:

Pets in Students' Families

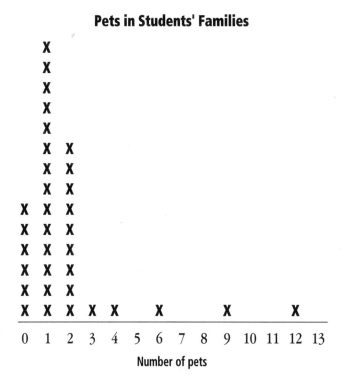

a. Would it be possible to have a data set for which the median number of pets for students is 0? Explain.

b. Would it be possible to have a data set for which the mean number of pets for students is 0? Explain.

Notebook Checklist

Journal Organization

_____ Problems and Mathematical Reflections are labeled and dated.

_____ Work is neat and easy to find and follow.

Vocabulary

_____ All words are listed.

_____ All words are defined or described.

Quizzes and Check-Ups

_____ Check-Up 1   _____ Quiz

_____ Check-Up 2

Homework Assignments

_____ _____

_____ _____

_____ _____

_____ _____

_____ _____

_____ _____

_____ _____

_____ _____

_____ _____

_____ _____

_____ _____

_____ _____

_____ _____

Self-Assessment

Vocabulary

Of the vocabulary words I defined or described in my journal, the word _____ best demonstrates my ability to give a clear definition or description.

Of the vocabulary words I defined or described in my journal, the word _____ best demonstrates my ability to use an example to help explain or describe an idea.

Mathematical Ideas

1. **a.** I learned these things about collecting, displaying, and analyzing data from *Data About Us:*

   b. Here are page numbers of journal entries that give evidence of what I have learned, along with descriptions of what each entry shows:

2. **a.** These are the mathematical ideas I am still struggling with:

   b. This is why I think these ideas are difficult for me:

   c. Here are page numbers of journal entries that give evidence of what I am struggling with, along with descriptions of what each entry shows:

Class Participation

I contributed to the classroom discussion and understanding of *Data About Us* when I . . . (Give examples.)

Answers to Check-Up 1

1. a. There are 25 people represented. The mode is 7. The median is 6. The range is from 3 letters to 9 letters.

   b. There are 36 people represented. The mode is November. The median and range are not appropriate representations for categorical data.

2. Students need to place at least one X over the 6 and one X over the 16. The remaining nine pieces of data must be distributed so that one or more Xs are above the 12 and so that, if you count Xs starting at either end, the sixth X is over the 12. Possible answers:

Name Lengths

```
              X     X                       X
        X     X     X     X  X  X  X  X
      _____
       5   6   7   8   9  10 11 12 13 14 15 16 17
                     Number of letters
```

Name Lengths

```
                          X
        X  X  X  X  X      X  X  X  X  X
      _____
       5   6   7   8   9  10 11 12 13 14 15 16 17
                     Number of letters
```

3. a. Possible answers: How many books do you read in a month? How many books do you buy in a year? How many days should the book fair run? What is the most you are willing to pay for a paperback book?

   b. Possible answers: What is your favorite subject for reading? Who is your favorite author? What should we sell at the book fair in addition to books?

Answers to Quiz

1–3. Answers will depend on the class's data.

4. a. (This data could also be expressed as a back-to-back stem plot.)

Grade 5 Data

```
13 | 8  8  8  9
14 | 1  2  4  6  7  7  7
15 | 0  0  1  1  1  1  2  2  2  2  3  3  5  5  6  6  7  8
16 |
17 | 1
```

Grade 8 Data

```
14 | 7
15 | 6 9
16 | 0 0 1 2 2 2 2 3 4 5 5 8 8 8 8 9
17 | 1 2 4 6
```

b. Answers should fall around 150 cm. Students may use one or more of these measures to justify their answer:

median: 151   modes: 151, 152

mean: 149.77   range: 138–171

c. Answers should fall around 164 cm. Students may use one or more of these measures to justify their answer:

median: 164   modes: 162, 168

mean: 164.43   range: 147–176

d. The data seem to cluster around one 10-centimeter range. In grade 5, it is the 150s; in grade 8, it is the 160s. There is an outlier on the tall end in grade 5 and somewhat of an outlier on the short end in grade 8. The distribution seems to shift up 10 centimeters from one grade to the next.

e. The median increases to 164.5; the modes remain the same; the mean increases from 164.43 to 165.73.

5.   a.

Frog Data

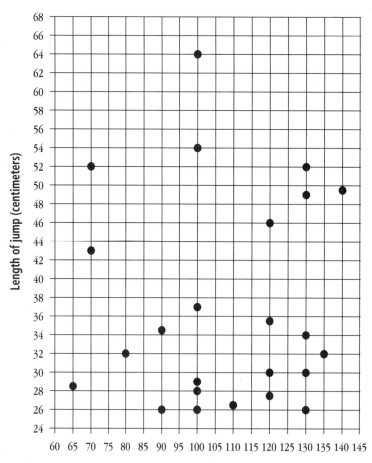

b. The graph shows a wide range in jump length for frogs with a mass of 100 grams—this is about the midpoint in the frogs' masses. More frogs were 100 grams than any other mass, so the mode is 100 grams. Jump lengths were fairly evenly distributed over the different masses.

c. Knowing the weight does not help you to predict how far the frog will jump.

d. 105 grams

e. 33 centimeters

f. 7 frogs

g. 6 frogs

Answers to Check-Up 2

1. a. 2

    b. 3

2. a. numerical

    b. 5–99

    c. 73; Only one student had that score.

3. a. 3 students

    b. 3 students

    c. The cuffs will be loose.

    d. The cuffs will be tight.

4. a. Possible answers: How old are you? How tall are you in inches? How many brothers and sisters do you have? How many times a month do you go to the movies?

    b. Possible answers: What is your favorite food? Movie? Book? Animal? Subject in school? What sports do you play, if any? What sports do you dislike, if any?

Answers to the Question Bank

1. There are 10 people represented. The mode is 4. The median is 5. The range is 4–7.

2. a. 22 paces

b. The line plot and the bar graph show representations of the same data. The bar graph requires a vertical scale to read the numbers of data in each group, while the numbers on the line plot can be found by counting the X's in each column.

Paces to the Gym

c. The student who traveled the distance in 25 paces has the shorter pace. It took that student more paces to travel the same distance as the student who traveled the distance in only 17 paces.

3. Possible answers:

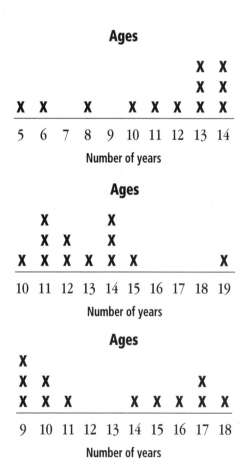

4. **a.** 30 children

   b. Possible answer: 4, 4, 5, 5, 6, 6

   c. No, they would not have to be the same. Other sets could be, for example, 2, 4, 5, 5, 6, 8 and 1, 4, 5, 5, 6, 9.

5. You are dividing the total number of children among many families. If each family has two children and there are some children left over, they must be divided equally among all the families. Since you are dealing with the distribution of numbers, this comes up mathematically as a fraction of a number. The mean of the distribution would be a whole number only if the number of families were a factor of the number of children.

6. **a.** Answers will vary. The statistic was probably determined by recording how far a sample group of people recorded walked in a day. The mean distance was calculated from the sample. This was multiplied by the number of days in a year, and the result was multiplied by the life expectancy of people today.

   b. Possible answer: I would have people wear a pedometer for a week and keep daily records. I would choose people of both genders and all age groups, cultures, and occupations. It would also be necessary to find the current life expectancy, which differs by gender and other factors. When I found the average distance walked for a day for my group, I would multiply it by the number of days in a year and the life expectancy.

7. **a.** yes; If there were more students without pets than students with pets, the median would be 0.

   b. yes; If no students had pets, the mean would be 0.

The final assessment for *Data About Us* is the Is Anyone Typical? Project. The project requires students to conduct a statistical investigation to determine some "typical" characteristics of students in their class or their school. As mentioned in the "Assigning the Project" section, each group can tackle their own complete investigation, or you can write a survey as a class and assign each group the task of analyzing and interpreting the results of one survey question.

When you assign the project remind students of the question they are trying to answer: "What are some characteristics of a typical middle-school student?" Remind students of some of the typical characteristics they have already determined, such as name length and height. Then, have a class discussion about the four steps involved in a statistical investigation.

Step 1. Posing Questions

Before writing their surveys, students must decide what information they want to know. An interesting survey will collect both categorical and numerical information. You will want to check each group's list of questions before allowing them to proceed with their survey. You need to be vigilant about the kinds of questions your students ask. Questions that might embarrass students should not be included. Two questions that may help students to focus are, What will you learn from this question? and, How will this question help you learn about the typical middle-school student? If you are writing the survey as a class, have students brainstorm about questions they might ask and then work together to narrow down the list.

After students have decided what they want to know, they need to make sure the questions they ask are precise and unambiguous. Have a discussion about the types of questions that work best as "fill in the blanks," such as

How old are you in months? _____

and the types of questions that work best if choices are provided. For example,

What do you do when you are bored? Check the one response that best describes you:
_____ watch TV   _____ listen to the radio
_____ read a book   _____ play with a pet
_____ talk on the telephone _____ complain

If students want to know someone's attitude about something, they can ask how strongly they feel about the question being asked. For example,

Circle only one answer: 1 means strongly disagree, and 5 means strongly agree.

Students should be allowed to wear hats in school?   1   2   3   4   5

Step 2. Collecting the Data

Students need to decide whether they will survey their class only or a larger group. The survey will need to be produced, duplicated, and distributed. You may want to coordinate the data collection among all of your classes so classes from which data are collected are not disturbed several times.

Step 3. Analyzing the Data

Once they have collected the data, students need to organize, tally, and display the data. They must decide which displays and which measures of centers are appropriate for each set of data.

Step 4. Interpreting the Results

After the data have been analyzed, students need to interpret the results and write a report or create a poster to display their findings. The report should include the questions asked, information about how the data were collected, appropriate data displays and measures of center, and concluding statements about what is typical about the data.

If you create the survey as a class, and assign one question to each group, you will want to come together again as a class to "pool" the results and assemble the characteristics of a typical student. The photo below shows how one teacher displayed the results of a class investigation.

Suggested Scoring Rubric

This rubric for scoring the project employs a scale that runs from 0 to 4, with a 4+ for work that goes beyond what has been asked for in some unique way. You may use this rubric as presented here or modify it to fit your district's requirements for evaluating and reporting students' work and understanding.

4+ Exemplary Response
- Complete, with clear, coherent explanations
- Shows understanding of the statistical concepts and procedures
- Satisfies all essential conditions of the problem and goes beyond what is asked for in some unique way

4 Complete Response
- Complete, with clear, coherent explanations
- Shows understanding of the statistical concepts and procedures
- Satisfies all essential conditions of the problem

3 Reasonably Complete Response
- Reasonably complete; may lack detail in explanations
- Shows understanding of most of the statistical concepts and procedures
- Satisfies most of the essential conditions of the problem

2 Partial Response
- Gives response; explanation may be unclear or lack detail
- Shows some understanding of some of the statistical concepts and procedures
- Satisfies some essential conditions of the problem

1 Inadequate Response
- Incomplete; explanation is insufficient or not understandable
- Shows little understanding of the statistical concepts and procedures
- Fails to address essential conditions of problem

0 No Attempt
- Irrelevant response
- Does not attempt a solution
- Does not address conditions of the problem

Samples of Student Projects

The samples that follow are from a class in which each group investigated one question. The first project investigates the states that students have visited. The second looks at the class's favorite radio station.

Sample 1

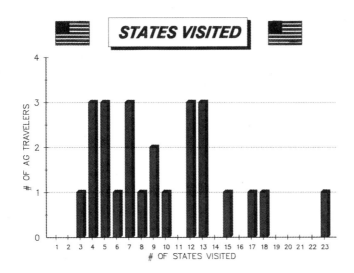

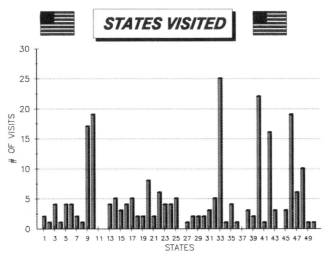

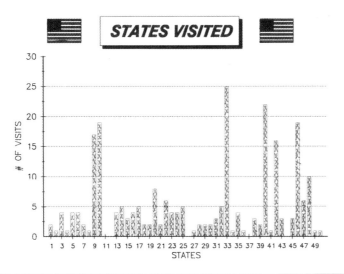

Key

1. Alabama ___
2. Alaska ___
3. Arizona ___
4. Arkansas ___
5. California ___
6. Colorado ___
7. Connecticut ___
8. Deleware ___
9. Flordia ___
10. Georgia ___
11. Hawaii ___
12. Idaho ___
13. Illinois ___
14. Indiana ___
15. Iowa ___
16. Kansas ___
17. Kentucky ___
18. Louisana ___
19. Maine ___
20. Maryland ___
21. Massachusetts
22. Michigan ___
23. Minnesota ___
24. Mississippi ___
25. Missouri ___

26. Montana ___
27. Nebraska ___
28. Nevada ___
29. New hampshire ___
30. New Jersey ___
31. New Mexico ___
32. New York ___
33. North Carolina ___
34. North Dakota ___
35. Ohio ___
36. Oklahoma ___
37. Oregon ___
38. Pennsylvania ___
39. Rhode Island ___
40. South Carolinia
41. South Dakota ___
42. Tennessee ___
43. Texas ___
44. Utah ___
45. Vermont ___
46. Virginia ___
47. Washington ___
48. West Virginia ___
49. Wisconsin ___
50. Wyoming ___

A Teacher's Comments

I evaluated each step of the investigation separately, using a four-point scale for each step. These students received 10 of 16 possible points for their project. I would recommend that they redo the "Interpret the Results" section of their project.

Step 1. Posing Questions (3 points)
These students asked interesting questions. However, they should have clarified what they meant by "been to." Did they mean visited or just traveled through?

Step 2. Collecting the Data (3 points)
The students did not mention the number of students they surveyed. I can determine the number of students that were surveyed from the first graph by counting and adding the heights of the bars ($1 + 3 + 3 + 1 + 3 + 1 + 2 + 1 + 3 + 3 + 1 + 1 + 1 + 1 = 25$).

Step 3. Analyzing the Data (3 points)
The graphs are quite interesting. Students explored both categorical and numerical data. The first graph shows the numbers of states students have visited. No summary statistics are given for this graph. It is interesting to note that the range in the number of states visited is 3 to 25 and the median is between 8 and 9 states.

The other two graphs show categorical data. The first graph is a bar graph. The students used numbers on the horizontal axis to represent the states and provided a key to the right of the graph. The second graph was intended to be a line plot, but has a vertical axis like a bar graph. This vertical axis is not necessary, since the numbers of Xs indicate the frequencies. The students determined that the mode state is 40, South Carolina. (These students live in North Carolina, and so did not consider it a "state visited.") These students, correctly, did not include any other statistics for these two graphs. The mean and median would are not appropriate measures for categorical data.

Step 4. Interpreting the Results (1 point)
The students provided only one summary statement about their results—that the mode state is South Carolina. They could have discussed the fact that the mode state and the states close to the mode—Virginia, Georgia, Florida, and Tennessee—are neighboring states of North Carolina. They might also have mentioned that, in the first graphs, values of 23, 14, 17, and 18 are unusual values, and explored why these students had visited so many states.

Sample 2

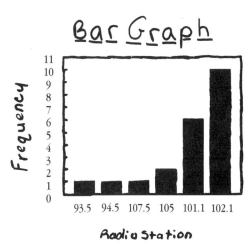

Bar Graph

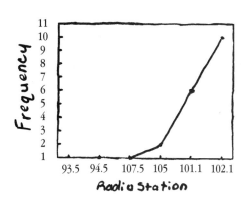

Pie Graph

93.5 (4.8%)

94.5 (4.8%)

107.5 (4.8%)

105 (9.5%)

102.1 (47.6%)

102.1 (28.6%)

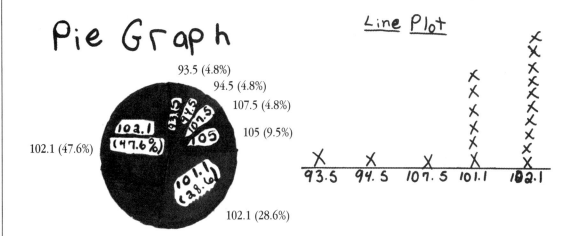

Line Plot

Bonnie and I have done a math project.
Our question is what is your favorite radio
station? The mode was 102 jams. We didn't have
a mean, median, or range because we're dealing
with station names. Above we have a pie graph,
bar graph, and two types of line plots. They may
look different, but they show the same kind of
data.
Ten people said htey liked 102 jams. Six people
said they liked country 101.1. One person said
they liked 107.5, one said they liked 94.5, and
one said they liked 93.5. We hade a total of
twenty-one students. That is our project.

A Teacher's Comments

I evaluated each step of the investigation separately, using a four-point scale for each step. These students received 13 of 16 possible points for their project.

Step 1. Posing Questions (4 points)
The question was clearly stated.

Step 2. Collecting the Data (3 points)
In their summary paragraphs, the students indicate that they surveyed 21 students, yet their graphs show data from only 19 students. Perhaps two students did not have a favorite station. Students should have mentioned this or included a "no favorite" category.

Step 3. Analyzing the Data (3 points)
These students used four different displays. The line plot, the bar graph, and the circle graph are appropriate. The line graph is not appropriate because it is designed to show change over time. On the line plot, the Xs are not aligned, so the stack of Xs for 102.1 looks shorter than it should relative to the stack for 101.1.

Step 4. Interpreting the Results (3 point)
These students included a summary statement. They acknowledge the inappropriateness of the mean, median, or range for categorical data. I would have liked them to talk a bit more about the radio stations. Stations 101.1 and 102.1 are far more popular than the other three choices.

Blackline Masters

Times and Distances to School

Numbers of Seeds in Pumpkins

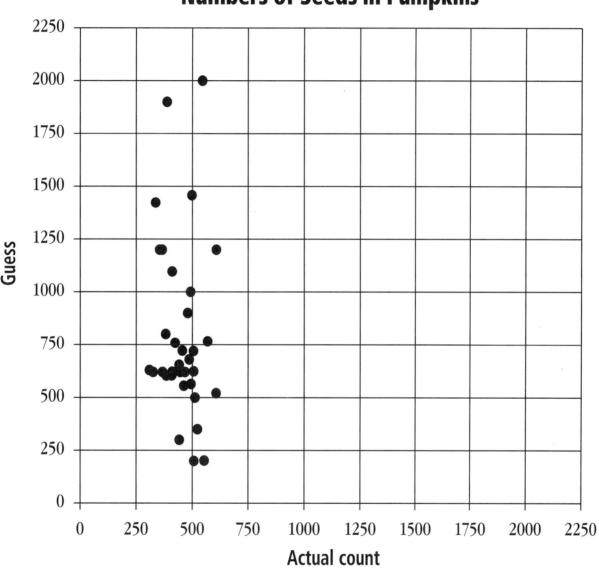

Gather data about the total number of letters in the first and last names of students in your class.

A. Find a way to organize the data so you can determine the typical name length.

B. Write some statements about your class data. Note any patterns you see.

C. What would you say is the typical name length for a student in your class?

D. If a new student joined your class today, what would you predict about the length of that student's name?

A. Write some statements about the name lengths for student's in Ms. Jeckle's class. Describe any interesting patterns you see in the data.

B. In what ways are the two graphs alike? In what ways are they different?

C. How does the data from Ms. Jeckle's class compare with the data from your class?

Name Lengths of Ms. Jeckle's Students

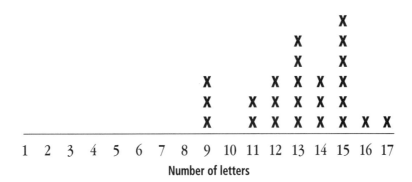

Name Lengths of Ms. Jeckle's Students

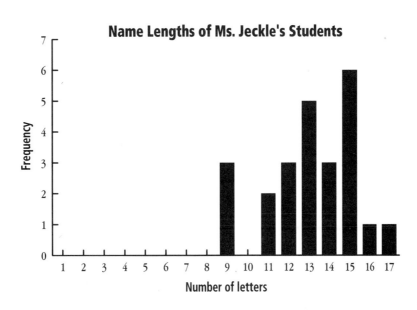

There are 15 students in a class. The mode of the name lengths for the class is 12 letters, and the range is from 8 letters to 16 letters.

A. Determine a set of name lengths that has this range and mode.

B. Make a line plot to display your data.

C. Use your line plot to help you describe the shape of your data. For example, your data may be bell-shaped, spread out in two or more clusters, or grouped together at one end of the graph.

Name	Letters
Jeffrey Piersonjones	19
Thomas Petes	11
Clarence Surless	15
Michelle Hughes	14
Shoshana White	13
Deborah Locke	12
Terry Van Bourgondien	19
Maxi Swanson	11
Tonya Stewart	12
Jorge Bastante	13
Richard Mudd	11
Joachim Caruso	13
Roberta Northcott	16
Tony Tung	8
Joshua Klein	11
Janice Vick	10
Bobby King	9
Jacquelyn McCallum	17
Kathleen Boylan	14
Peter Juliano	12
Linora Haynes	12

Class Name Lengths

```
             X  X
             X  X  X
             X  X  X  X              X
    X  X  X  X  X  X  X  X  X  X      X
   ─────────────────────────────────────
    7  8  9  10 11 12 13 14 15 16 17 18 19 20
             Number of letters
```

Cut a strip of 21 squares from a sheet of grid paper. Write the Michigan class's name lengths in order from smallest to largest on the grid paper as shown here.

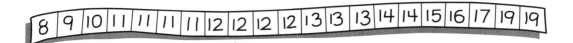

Now, put the ends together and fold the strip in half.

A. Where does the crease land? How many numbers are to the left of the crease? How many numbers are to the right of the crease?

Suppose a new student, Suzanne Mannerstrale, joins the Michigan class. The class now has 22 students. On a strip of 22 squares, list the name lengths, including Suzanne's, in order from smallest to largest. Fold this strip in half.

B. Where is the crease? How many numbers are to the left of the crease? How many numbers are to the right of the crease?

Experiment with your cards to see if you can perform each task described below. Keep a record of the things you try and the discoveries you make.

A. Remove two names without changing the median.

B. Remove two names so the median increases.

C. Remove two names so the median decreases.

D. Add two new names so the median increases.

E. Add two new names so the median decreases.

F. Add two new names without changing the median.

Think of some things you would like to know more about. Then, develop some questions you could ask to gather information about those things.

A. Write two questions that have categorical data as answers.

B. Write two questions that have numerical data as answers.

Favorite Kinds of Pets

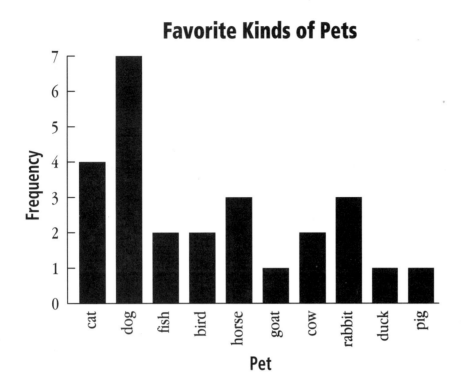

Numbers of Pets

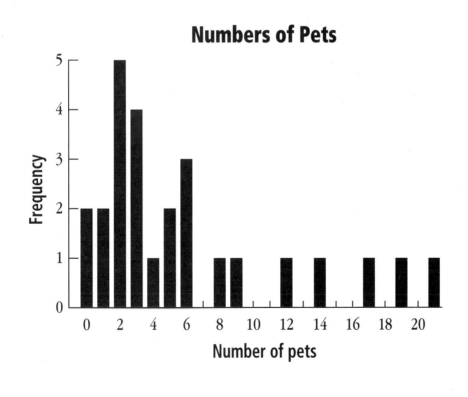

Decide whether each question below can be answered by using data from the graphs the students created. If a question can be answered, give the answer and explain how you got it. If a question cannot be answered, explain why not and tell what additional information you would need to answer the question.

A. Which graph shows categorical data, and which graph shows numerical data?

B. What is the total number of pets the students have?

C. What is the greatest number of pets that any student in the class has?

D. How many students are in the class?

E. How many students chose cats as their favorite kind of pet?

F. How many cats do students have as pets?

G. What is the mode for the favorite kind of pet?

H. What is the median number of pets students have?

I. What is the range of the numbers of pets students have?

J. Tomas is a student in this class. How many pets does he have?

K. Do the girls have more pets than the boys?

Stem plot of unsorted data:

```
0 | 8 8 5 5 5 6
1 | 5 5 5 5 9 5 5 7 5 0 5 0 5 1 7 0 0
2 | 2 5 0 5 0 0 0 0 0 1
3 | 0 5 0 0 5
4 |
5 | 0
6 | 0
```

Stem plot of sorted data:

Travel Times to School (minutes)

```
0 | 5 5 5 6 8 8
1 | 0 0 0 0 1 5 5 5 5 5 5 5 5 5 7 7 9
2 | 0 0 0 0 0 0 1 2 5 5
3 | 0 0 0 5 5
4 |
5 | 0
6 | 0
```

Key

2 | 5 means 25 minutes.

Read "Making a Stem-and-Leaf Plot" to explore how to make a stem-and-leaf plot of the travel-time data. After you have completed your stem plot of the data, answer these questions.

A. Which students probably get to sleep the latest in the morning? Why do you think this?

B. Which students probably get up earliest? Why do you think this?

C. What is the typical time it takes for these students to travel to school?

Which class did better overall in the jump-rope activity? Use what you know about statistics to help you justify your answer.

Numbers of Jumps

Ms. R's class								Mr. K's class							
8	7	7	7	5	1	1	0	1	1	2	3	4	5	8	8
				6	1	1	1	0	7						
	9	7	6	3	0	0	2	3	7	8					
					5	3	3	0	3	5					
					5	0	4	2	7	8					
							5	0	2	3					
						2	6	0	8						
							7								
				9	8	0	8								
				6	3	1	9								
							10	2	4						
						3	11								
							12								
							13								
							14								
							15	1							
							16	0	0						
							17								
							18								
							19								
							20								
							21								
							22								
							23								
							24								
							25								
							26								
							27								
							28								
							29								
							30	0							

Key

7 | 3 | 0 means 37 jumps for
Ms. R's class and
30 jumps for Mr. K's class.

Initials	Height (inches)	Arm span (inches)
NY	63	60
JJ	69	67
CM	73	75
PL	77	77
BP	64	65
AS	67	64
KR	57	57

Height and Arm Span Measurements

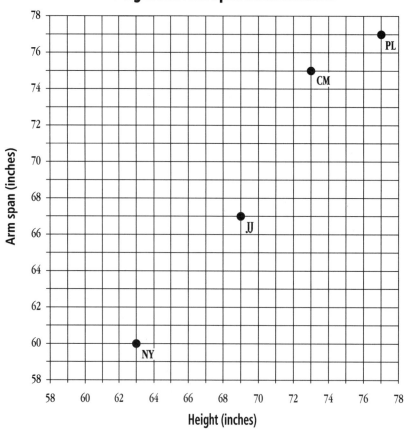

Think about this question: If you know the measure of a person's arm span, do you know anything about his or her height?

To help you answer this question, you will need to collect some data. With your class, collect the height and arm span of each person in your class. Make a coordinate graph of your data. Then, use your graph to answer the question above.

Study the graph below, which was made using the data from Problem 3.1.

A. Look back at the data on page 31. On Labsheet 4.2, locate and label with initials the points for the first five students in the table.

B. If you know how long it takes a particular student to travel to school, can you know anything about that student's distance from school? Use the graph to help you answer this question. Write a justification for your answer.

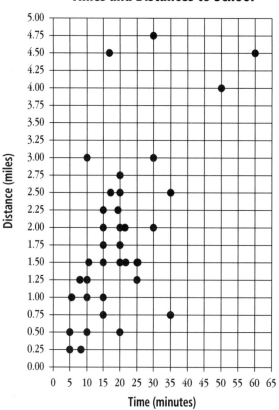

Times and Distances to School

Numbers of People in Households

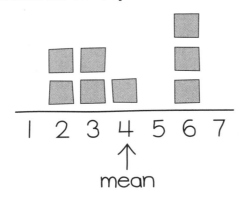

Numbers of People in Households

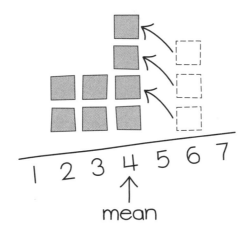

Subtract a total of 6 cubes:

$6 - 2 = 4$
$6 - 2 = 4$
$6 - 2 = 4$

Numbers of People in Households

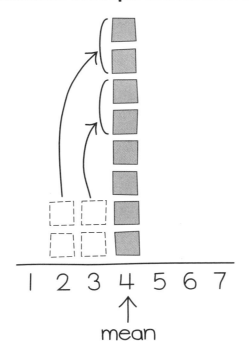

Add a total of 6 cubes:

$3 + 1 = 4$
$3 + 1 = 4$
$2 + 2 = 4$
$2 + 2 = 4$

What are some ways to determine the average number of people in these eight households?

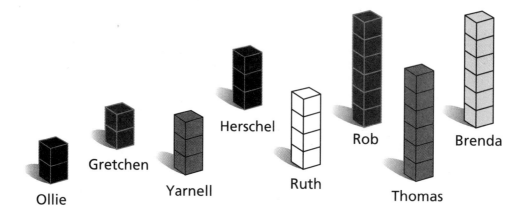

A. Make a set of cube towers to show the size of each household.

B. Make a line plot of the data.

C. How many people are there in the eight households altogether? Describe how you determined your answer.

D. What is the mean number of people in the eight households? Describe how you determined your answer.

Name	Number of people in household
Geoffrey	6
Betty	4
Brendan	3
Oprah	4
Yancey	2
Reilly	3
Tonisha	4
Barker	6

A. Try to find two more sets of eight households with a mean of 4 people. Use cubes to show each set, and then make line plots that show the information from the cubes.

B. Try to find two different sets of nine households with a mean of 4 people. Use cubes to show each set, and then make line plots to show the information from the cubes.

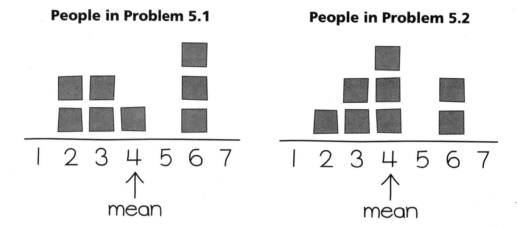

People in Problem 5.1   **People in Problem 5.2**

A. Using the definition from the United States census, how many people are in your household?

B. Collect household data from everyone in your class, and make a display to show the information.

C. What is the mean number of people in your class's households? Describe how you determined your answer.

A. Look at the table on page 60, and complete these statements. You may want to use your calculator.

The total number of students is _____.

The total number of movies watched is _____.

The mean number of movies watched is _____.

B. Data is added for Lucia, who watched 42 movies last month. This value is an outlier. How does the stem plot change when this value is added? What is the new mean? Compare the mean from part A to the mean after this value is added. What do you notice?

C. Data is added for Tamara, who was home last month with a broken leg. She watched 96 movies. What is the mean of the data now? Compare the means you found in parts A and B with this new mean. What do you notice? Why?

D. Data for eight more students are added:

Tommy	5	Robbie	4
Alexandra	5	Ana	4
Kesh	5	Alisha	2
Kirsten	5	Brian	2

These data are not outliers, but now there are several students who watched only a few movies in one month. What is the mean of the data now? Compare the means you found in parts A, B, and C with this new mean. What do you notice? Why?

Dear Family,

This next unit in your child's course of study in mathematics class this year is *Data About Us.* Its focus is data investigation, and it teaches students to organize, display, analyze, and interpret data. Your child will learn to create and interpret many different types of data displays and to compute statistics and use them to help describe data.

Data About Us is placed early in the school year to help students get to know one another. The unit provides opportunities for students to ask questions about themselves, and then to collect data to help answer these questions. Students explore the lengths of their names, the distances they live from school, the numbers of times they can jump rope, the number of pets they have, their heights, and the lengths of their left feet.

Your child will learn to create line plots, bar graphs, scatter plots, and stem-and-leaf plots and to recognize and interpret patterns shown in these displays. Your child will also learn to compute the mode, median, mean, and range of a data set and to use these statistics to describe data and to make predictions.

You can help your child in several ways:

- Look with your child for uses of data in magazines, newspapers, and on TV.

- Point out examples of graphical displays and ask your child questions about the information shown.

- Ask your child about the data studied in class. What were the typical (mode, median, or mean) values for these data?

- Look over your child's homework and make sure all questions are answered and that explanations are clear.

As always, if you have any questions or concerns about this unit or your child's progress in the class, please feel free to call. All of us here are interested in your child and want to be sure that this year's mathematics experiences are enjoyable and promotoe a firm understanding of mathematics.

Sincerely,

axis, axes (page 43) The number lines that are used to make a graph. There are usually two axes perpendicular to each other (see *bar graphs* or *coordinate graphs* for examples). The vertical axis is called the *y*-axis and the horizontal axis is called the *x*-axis.

bar graph (bar chart) (page 8) A graphical representation of a table of (discrete or "counted") data in which the height of each bar indicates its value or frequency. The bars are separated from each other to highlight that the data are discrete or "counted" data. The horizontal axis shows the values or categories and the vertical axis shows the frequency or tally for each of the values or categories on the horizontal axis. Bar graphs may be used to display categorical or numerical data.

Favorite Colors

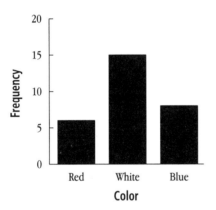

categorical data (page 22) Values that are "words" that represent possible responses within a given category. Frequency counts can be made of the values for a given category. As an example, see *survey* and the examples that follow.

- Months of the year in which people have birthdays (values may be January, February, March, and so on)

- Favorite color to wear for a t-shirt (values may be magenta, Carolina blue, yellow, and so on)

- Kinds of pets people have (values may be cats, dogs, fish, horses, boa constrictors, and so on)

coordinate graph (page 43) A graphical representation of pairs of related numerical values. The data are sorted into pairs of numbers with each pair associated with one person (for example, height and arm span of each person measured) or object (for example, length and width of different-size rectangles). One axis is designated to show one value of each pair (for example, height on the horizontal axis) and the other axis shows the other value of each pair (for example, arm span on the vertical axis). The graph below shows a partially complete representation of the data in the table.

Measures (inches)		
Initials	Height	Arm span
JJ	69	67
KR	57	57
NY	63	60
AS	67	64
BP	64	65
CM	73	75
PL	77	77

Height and Arm Span Measurements

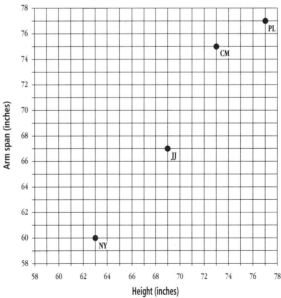

data (page 3) Values such as counts, ratings, measurements, or opinions that are gathered to answer questions. This table shows data that are mean temperatures in three cities.

Daily Mean Temperatures	
City	**Mean temp.**
Mobile, AL	67.5° F
Boston, MA	51.3° F
Spokane, WA	47.3° F

line plot (page 7) A quick, simple way to organize data along a number line where the Xs (or other symbols) above a number represent the frequency tally of data for that value of the data.

Numbers of Siblings Students Have

```
          X
X      X  X      X
X  X   X  X   X  X
X  X   X  X   X  X   X      X
0  1   2  3   4  5   6  7   8
        Number of siblings
```

mean (page 55) Of a distribution, a value calculated from the data. It can be thought of as the value that a set of data have if all the data are the same value. For example, the mean number of siblings for the above distribution is $21\frac{18}{19}$ siblings, or about 3 siblings. If all 19 families had the same number of siblings, they would each have about 3 siblings.

median (page 12) Of a distribution, the numerical value that marks the middle of an ordered set of data. Half the data occur above the median, and half the data occur below the median. The median of the distribution of siblings is 3 because the tenth (middle) value in the ordered set of 19 values (0, 0,

0, 1, 1, 2, 2, 2, 3, 3, 3, 4, 4, 5, 5, 5, 6, 8) is 3 siblings.

mode (page 9) Of a distribution, the category or numerical value that occurs most often. For example, the mode of the distribution of the number of siblings is 2. It is possible to have more than one mode; we talk about data that are bimodal (2 modes) and trimodal (3 modes).

numerical data (page 22) Values that are numbers such as counts, measurements, and ratings. As an example, see *data* and the examples that follow.

- Numbers of children in families
- Pulse rates indicating how many heart beats occur in a minute
- Height
- How much time people spend reading in one day
- How much people value something, such as: On a scale of 1 to 5 with 1 as "low interest," how would you rate your interest in participating in the school's field day?

outlier (page 36) One or more values that lie "outside" of a distribution of the data. An outlier is a value that may be questioned because it is unusual or because there may have been an error in recording or reporting the data. The 8 siblings in the example above may be considered an outlier. It is unusual for people to have that many siblings.

range (page 9) The range of a distribution is computed by stating the lowest and highest values. For example, the range of the number of siblings is from 0 to 8 people. Less frequently, the range is computed by finding the difference between the lowest and highest values.

scale (page 52c) The size of the unit used to calibrate the vertical axis number line (and the horizontal axis number line when data are numerical) of a plot or graph. For instance, the vertical axis in the example on the previous page represents the number of students. Each tick-mark represents 5 students. The labels for the tick marks at 10 students and 20 students are also shown in this example.

stem-and-leaf plot (stem plot) (page 32) A quick way to picture the shape of a distribution while including the actual numerical values in the graph. The *stem* of the plot is a vertical number line that represents a range of data values in a specified interval. The *leaves* are the numbers that are attached to the particular stem values. For example, the tens digit is indicated on the vertical axis as the stem, and the units digits next to the tens digit show the leaves that belong to that stem.

Numbers of Movies Seen Over the Summer

```
0 |
1 | 5 5 5 5
2 | 2 5 0
3 | 0 5
4 |
5 |
6 | 0
```

Back-to-back stem plots may be used to compare two sets of the same kind of data. The units for one set of data are placed on one side of the stem, and the units for the other set are placed on the other side.

survey (page 3) A method for data collection that usually employs written answers or interviews. Surveys ask one or more questions seeking such information as facts, opinions, or beliefs.

Favorite Colors	
Color	**Number of students**
Red	6
White	15
Blue	9

table (page 21d) A tool for organizing information in rows and columns. Tables let you list categories or values and then tally the occurrences. For an example, see *data*.